FRACTAL ANALYSIS
Basic Concepts and Applications

Series on Advances in Mathematics for Applied Sciences – Vol. 91

FRACTAL ANALYSIS
Basic Concepts and Applications

Carlo Cattani
University of Tuscia, Italy

Anouar Ben Mabrouk
University of Monastir, Tunisia & University of Kairouan, Tunisia
University of Tabuk, Saudi Arabia

Sabrine Arfaoui
University of Monastir, Tunisia
University of Tabuk, Saudi Arabia

World Scientific

N JERSEY · LONDON · SINGAPORE · BEIJING · SHANGHAI · HONG KONG · TAIPEI · CHENNAI · TOKYO

Published by

World Scientific Publishing Co. Pte. Ltd.
5 Toh Tuck Link, Singapore 596224
USA office: 27 Warren Street, Suite 401-402, Hackensack, NJ 07601
UK office: 57 Shelton Street, Covent Garden, London WC2H 9HE

Library of Congress Cataloging-in-Publication Data
Names: Cattani, Carlo, 1954–   author. | Ben Mabrouk, Anouar, author. | Arfaoui, Sabrine, author.
Title: Fractal analysis : basic concepts and applications / Carlo Cattani, University of Tuscia, Italy,
    Anouar Ben Mabrouk, University of Monastir, Tunisia &
    University of Kairouan, Tunisia, University of Tabuk, Saudi Arabia,
    Sabrine Arfaoui, University of Monastir, Tunisia, University of Tabuk, Saudi Arabia.
Description: New Jersey : World Scientific, [2022] | Series: Series on advances in mathematics for
    applied sciences, 1793-0901 | Includes bibliographical references and index.
Identifiers: LCCN 2021054614 | ISBN 9789811239434 (hardcover) |
    ISBN 9789811239441 (ebook) | ISBN 9789811239458 (ebook other)
Subjects: LCSH: Fractal analysis.
Classification: LCC QA614.86 .C38 2022 | DDC 514/.742--dc23/eng/20220104
LC record available at https://lccn.loc.gov/2021054614

British Library Cataloguing-in-Publication Data
A catalogue record for this book is available from the British Library.

For any available supplementary material, please visit
https://www.worldscientific.com/worldscibooks/10.1142/12345#t=suppl

Desk Editors: Anthony Alexander/Lai Fun Kwong

Typeset by Stallion Press
Email: enquiries@stallionpress.com

To our mothers, fathers, and families, to our teachers and everyone who contributed to the success of this work

Preface

Fractals may be observed almost everywhere in nature, and our daily life. They are observed in trees, rocks, snow, human body, cities, fractal smart grids ... etc. This makes them attractive subjects in scientific researches in pure and applied fields.

Fractals are also good models for the resolution of many problems in the pure mathematical point of view such as PDE, fractal domains and boundaries, dynamical systems, and turbulence modeling.

In the applied point of view, fractals are applied in prices modeling, market indices, climate factors, very fine technology, smart cities, traffic, arts, design, nature, and bio domains.

These facts have made them very interesting subjects in scientific researches, yielding a growing set of documentation about them.

However, there still exists a necessity of more references to be developed, especially due to the restriction of the majority of books dealing with fractal analysis and geometry to specific community of readers.

In the present volume, we have tried to elaborate a self-containing reference in the field of fractal analysis to be adapted for a large community of readers by providing the necessary developments.

Further, in the present volume, basic concepts of measure theory as a main tool in fractal analysis will be recalled with the necessary developments. Besides, some concepts from stochastic calculus such as martingales will be provided with detailed proofs due to the complexity of such concepts.

Next, the essential target of the book concerning the fractal dimensions, measures, sets are developed in a series of chapters, starting from the original definitions of Hausdorff measure and dimension, packing measures and dimension, the concept of capacity of sets and its relation to fractal measures and dimensions. Besides, the notion of multifractal formalism is also investigated with necessary developments, examples, and exercises.

Finally, the volume is completed with some applications of the concepts exposed previously.

To finish with this preface, we stress on the fact that the present volume, in fact, stems from many papers, lectures, presentations, and also discussions on the theory of fractal analysis, and its applications. The ideas gathered here are some times re-developed, improved, and completed with necessary developments. These may not be perfect and need sometimes to be criticized, corrected, and improved by readers. Any comments are welcome at any time. To participate in explaining and understanding the theory of fractal analysis, especially for beginners, some exercises and applications, that are simple to handle, are provided.

We hope that the present volume provides a basic and self-contained introduction to the ideas underpinning fractal analysis, and its related fields, especially for master's degree students, and PhD researchers. It may also serve for scientists and research in industrial sectors where the understanding and the construction of models for real world problems exhibits fractals. Recall that such concepts are now widely applied in finance, medicine, engineering, transport, images, signals, etc. This makes the present volume of interest for practitioners and theorists in these fields.

We would like finally to thank all persons without whose help the present work could not be realized. We would like to thank World Scientific Publishing Co. for the opportunity to realize our project.

About the Authors

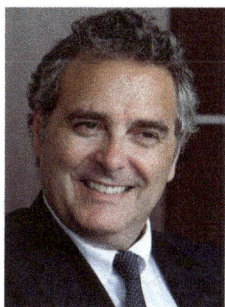

Carlo Cattani is currently Professor of Mathematical Physics and Applied Mathematics at the Engineering School (DEIM) of University of Tuscia. His scientific interests include, but are not limited to, wavelets, dynamical systems, fractals, fractional calculus, numerical methods, number theory, stochastic integro-differential equations, competition models, time-series analysis, nonlinear analysis, complexity of living systems, pattern analysis, computational biology, biophysics, history of science. He is the author (and co-author) of more than 300 scientific articles in international journals and many books.

Anouar Ben Mabrouk is qualified as Professor of Mathematics, and Mathematics and Applications, from the French ministry of Higher Education. He is an Associate Professor of Mathematics at the University of Kairouan, Tunisia, and a member of the Laboratory of Algebra, Number Theory, and Nonlinear Analysis, Faculty of Sciences, University of Monastir, Tunisia, and the Department of Mathematics, Faculty of Sciences, Tabuk University, Saudi Arabia. His main focus is on wavelets, fractals, probability/statistics, PDEs and related fields such as financial mathematics, time series, image/signal processing, numerical and theoretical aspects of PDEs. Currently, he is Professor of Mathematics at the University of Tabuk, Saudi Arabia.

Sabrine Arfaoui is Assistant Professor of Mathematics at the Laboratory of Algebra, Number Theory, and Nonlinear Analysis, University of Monastir, Faculty of Sciences, and the Department of Mathematics, Faculty of Sciences, Tabuk University, Saudi Arabia. Her main focus is on wavelet harmonic analysis, especially in the Clifford algebra/analysis framework and their applications in other fields such as fractals, PDEs, bio-signals/bio-images. Currently, Dr. Arfaoui is attached to University of Tabuk, Saudi Arabia, in a technical cooperation project.

Contents

List of Figures

List of Table

Chapter 1

Introduction

Fractals in analysis, geometry, and generally in science are nowadays very popular concepts. The appearance of such concepts in science is due essentially to Mandelbrot by pioneering their application in scientific modeling starting from financial time series issued from cotton price to a wide variety of applications in turbulence, plants, coastlines, galaxies, etc.

Indeed, in pure mathematics, fractals appeared in many contexts, such as the resolution of PDEs on fractal domains and PDEs with fractal boundaries ([Alimohammady et al (2017); Baleanu et al (2014); Cattani (2020); Xiaojun et al (2013); Xu et al (2014); Yang, Cattani et al (2014); Yang, Cattani and Xie (2015); Yang, Machado et al (2017); Yang, Srivastava et al (2015); Yang, Zhang et al (2014a,b); Yan et al (2014); Zhang, Cattani and Yang (2015)]). They may be also met in dynamical systems [Gervais (2009); Pesin (1997); Pesin and Climenhaga (2009)], and turbulence modeling [Benzi et al (1984); Cattani and Pierro (2013); Cattani (2010b); Frisch and Parisi (1985); Mandelbrot (1974)]. Since their appearance, fractals are also applied to explain the movement of prices, markets, financial indices, [Azizieh (2002); Benaych-Georges (2009); Ben Mabrouk, Ben Abdallah and Hamrita (2011); Calvet and Fisher (2008); Fan et al (2019); Fernandez-Martinez et al (2019); Fillol (2005); Hudson and Mandelbrot (2005); Mandelbrot (1997); Mandelbrot and Hudson (2006); Walter (2001)]. Besides, fractals are also applied for modeling instruments, such as fractal antenna ([Anguera et al (2020)], climate factors, and geophysical targets [Bozkus et al (2020); Chu (1999); Figueiredo et al (2014); Scholz and Mandelbrot (1989)]. In bio domains, such as medical applications, fractals are nowadays famous models [Badea et al, 2013; Castiglioni-Faini, 2019; Cattani, Pierro and Altieri, 2012; Karaca and Cattani, 2017; Mauroy et al, 2004; Sapoval and Filoche, 2010; Weibel, 1963].

Finally, we may also find them in signal/image processing, such as [Barnsley et al (1988); Cattani and Ciancio (2016); Chen, Cattani and Zhong (2014); Li et al (2014); Liu et al (2019); Vehel and Legrand (2004); Vehel and Vojak (1998)], traffic, arts, design, nature, etc. [Briggs (1992); Flake (1998); Li, Zhao and Cattani (2013); Mandelbrot (1982); Novak (2004); Pickover (1995); Scheuring and Riedi (1994); Wang et al (2014)].

Fractals have emerged in pure mathematics and pure mathematical analysis, and geometry has been developed as part and/or extension of measure theory, dimension theory, dynamical systems, and also non-Euclidean geometry. Mathematically, a Fractal set according to Mandelbrot derived its definition from the latin word fractus, which means in English 'broken', which explains the idea of characterizing next fractals by means of their Hausdorff dimension; generally and/or preferably, a non-integer real number quantity, which needs to be different from the topological dimension. Many variants have been developed by mathematicians to investigate such a notion. See the big list of references [Aversa and Bandt (1990); Baker and Schmidt (1970); Baker (1976); Barlow and Taylor (1992); Barral et al (2003); Batakis and Heurteaux (2002); Bélair (1987); Ben Nasr (1994, 1997); Ben Nasr and Bhouri (1997); Bernay (1975); Billingsley (1960, 1961); Billingsley and Henningsen (1975); Buck (1970, 1973); Cajar (1981); Cawley and Mauldin (1992); Cole (2000); Cole and Olsen (2003); Colebrook (1970); Collet et al (1987); Dai (1995); Dai and Liu (2008); David and Semmes (1997); Debussche (1998); Dryakhlov and Tempelman (2001); Edgar (1998, 2008); Eggleston (1949); Falconer (1994, 1985, 1990); Gurevich and Tempelman (1999); Heurteaux (1998); Lucas (2000); Makarov (1985); Mandelbrot (1993, 1995); Mattila (1995); Ngai (1997); Olivier (1998); Pesin (1997); Pesin and Climenhaga (2009); Riedi and Scheuring (1997); Rogers (1970); Selezneff (2011); Spear (1992); Taylor (1995); Veerman (1998); Wu (1998, 2005); Zeng et al (2012); Zhou and Feng (2011); Zhu and Zhou (2014)].

One of the simple ideas 'to understand' the concept of fractal dimensions, especially for non mathematicians, is to compare with the classical concepts of size (or the Lebesgue measure) when enforcing a scale change to the object in hand. For example, given an interval, or a segment $I = [a, b]$, its Lebesgue measure (length) is obviously $\ell(I) = b - a$. Multiplying I by a scale $\lambda \in \mathcal{R}^*$, for example, results in an interval λI whose Lebesgue measure is

$$\ell(\lambda I) = |\lambda|(b - a) = |\lambda|^1 \ell(I).$$

In 2-dimensional Euclidean space, a rectangle $I = [a, b] \times [c, d]$ has the Lebesgue measure (surface) $\ell(I) = (b - a)(d - c)$. Multiplying I by a scale

$\lambda \in \mathcal{R}^*$, as previously, results in a rectangle λI whom Lebesgue measure is

$$\ell(\lambda I) = |\lambda|^2(b-a)(d-c) = |\lambda|^2\ell(I).$$

Generally, in the n-dimensional Euclidean space, a cube $I = \prod_{i=1}^n[a_i, b_i]$ has the Lebesgue measure $\ell(I) = \prod_{i=1}^n(b_i - a_i)$. Multiplying I by a scale $\lambda \in \mathcal{R}^*$ results in a cube λI whom Lebesgue measure is

$$\ell(\lambda I) = |\lambda|^n \prod_{i=1}^{n}(b_i - a_i) = |\lambda|^n\ell(I).$$

The notion of dimension in the fractal analysis, and geometry aims to generalize this concept to general variants of dimensions by seeking suitable measures μ, which associates to the set λI an analogue quantity $|\lambda|^s\mu(I)$ for some generally non-integer s. This number will be the fractal dimension of the set I.

Many forms of measures have been introduced in the mathematical/physical literature to investigate this notion, such as, the Hausdorff, Caratheodory, packing, . . . , etc. Each of the introduced measures has been related to a notion of dimension, which describes the aim above. These measures, and dimensions have been next applied to re-define the concept of fractal sets, by imposing some equality on the different dimensions, and formalisms. The majority of formalisms are valid for Cantor type sets. In the general case, the concept of being Fractal or not is still under study, and needs more considerations.

One of the main problems behind the ambiguities is the fact that the computation of the fractal dimensions introduced, such as, the Hausdorff one is its impossibility, in the majority of cases, to be evaluated directly from the mathematical definition. We always have to come back to comparisons with Cantor types, self-similar, and scaling law. In some cases, we need to pass from the set to functional theory, and study instead of a set an associated function, which is generally irregular, and vice-versa. These facts, and difficulties are the main drawbacks, especially, in the applied fields when fractal models are investigated.

Indeed, from the applied point of view, nowadays, fractals may be observed almost everywhere in nature, and our daily life. They are observed in trees, rocks, snow, human body, cities, Fractal smart grids . . . etc. This makes them attractive subjects in scientific researches in pure, and applied fields.

Since their introduction in the early works of Mandelbrot, the documentation on fractals and their applications have been growing widely. Although, with the rapid developments in technology, and the appearance of many new natural phenomena, there still exists a great need to understand such new problems and their modeling with fractals. The majority of books dealing with fractal analysis and geometry are somehow restricted to specific readers, and, in the majority, addressed to a pure mathematical community. This is comprehensive, and may be explained by the need of the developments of the original, and the first theoretical steps, and basic concepts for the theory.

Next, with the dramatic development in technology, and science, fractal analysis has become a need in academic studies, especially, for young researchers. Consequently, the scientific academic community has had a great need to develop references in different forms, such as self-containing references, which respond mostly to the questions of young readers in Master's and PhD levels. The present volume will be part of these self-containing references, and will help readers to win a lot of time by providing the necessary developments.

This book is composed of nine chapters. A first introductory one in which a literal introduction is developed discussing the topic generally. Chapter 2 is concerned with the presentation of the basic concepts of measure theory, which constitutes the main tool and the main prerequisite for fractal analysis. Original developments of measure theory are provided with the necessary details.

Chapter 3 is concerned with the study of the notion of martingales, especially in the discrete case. This chapter constitutes a part of the useful concepts applied in fractal analysis, and thus responds to our aim to provide a self-contained reference in fractal analysis. The essential results to be applied in the nest chapters are recalled with necessary details.

Chapter 4 constitutes the starting step in the target theory to be developed in the present volume. Indeed, this chapter is the basic tool and the first step in fractal analysis and fractal geometry, as well as any field applying these notions. Basic construction of the so-called Hausdorff measure and the associated Hausdorff dimension will be exposed in detail.

Chapter 5 is devoted to the presentation of another variant of fractal dimensions, the so-called capacity dimension of sets, which coincides in some cases with the Hausdorff dimension.

In Chapter 6, the notion of the packing measure and packing dimension are exposed as other variants of fractal measures and dimensions.

Chapter 7 is devoted to the presentation of the multifractal formalism for measures. In this chapter, we essentially focus on the simple case of multifractal measures where the construction of associated Gibbs ones is always possible. It is shown that the possibility of constructing Gibbs measures on the supports of the multifractal measures is the essential cause of the validity of the multifractal formalism. The first developments in the same direction of the subject presented in this chapter are due to [Brown et al (1992); Peyriere (1992)]. Besides, for more details, and more examples, and cases, readers may refer also to [Ben Mabrouk (2008a,b,c); Ben Mabrouk and Aouidi (2011, 2012); Ben Mabrouk, Aouidi and Ben Slimane (2014, 2016); Ben Nasr (1994, 1997); Ben Nasr and Bhouri (1997); Ben Nasr, Bhouri and Heurteaux (2002); Billingsley (1960, 1961, 1965, 1968); Billingsley and Henningsen (1975); Brown et al (1992); David and Semmes (1997); Edgar (1998, 2008); Falconer (1994, 1990); Heurteaux (1998); Mandelbrot (1995); Menceur, Ben Mabrouk and Betina (2016); Menceur and Ben Mabrouk (2019); Mignot (1998); Ngai (1997); Olsen (1995); Pesin (1997); Pesin and Climenhaga (2009); Peyriere (1992); Riedi (1995); Rogers (1970); Spear (1992); Wu (1998)].

Chapter 8 is devoted to the presentation of the multifractal formalism for measures. It is concerned with the exposition of the multifractal extensions of the notions exposed in the previous chapters. Multifractal generalizations of Hausdorff and packing measures as well as their associated dimensions are exposed. We recall in this context that this chapter is essentially based on Olsen's works in [Cole and Olsen (2003); Cole (2000); Olsen (1995, 1996, 2000); O'Neil (1997)]. However, more information, developments, and extensions may also be found in [Ben Mabrouk (2008a,b,c); Ben Mabrouk and Aouidi (2011, 2012); Ben Mabrouk, Aouidi and Ben Slimane (2014, 2016); Ben Nasr (1994, 1997); Ben Nasr and Bhouri (1997); Ben Nasr, Bhouri and Heurteaux (2002); Menceur, Ben Mabrouk and Betina (2016); Menceur and Ben Mabrouk (2019)].

Finally, Chapter 9 is concerned with the development of some applications of the concepts exposed in the previous chapters, in order to show the utility of the notion of the fractal analysis, on the one hand, and to provide the community of non mathematicians with some ideas of concrete applications, on the other.

Chapter 2

Basics of Measure Theory

We propose in this chapter to recall some basic concepts of measure theory that will be applied later. We will try to review with some details all the concepts needed from measure theory, and focus on the most closed concepts to fractal analysis, and geometry, which constitute the principal goal of the book. Readers may refer to [David and Semmes (1997); Edgar (1998, 2008); Falconer (1990); Lambert (2012); Robert (2020); Yeh (2014)].

2.1 σ-algebras

Along the present chapter, $\mathcal{B}$ will designate a σ-algebra on the Euclidean space $\mathbb{R}^d$, $d \in \mathbb{N}$ is an integer.

Definition 2.1. Let X be a set, and $\mathcal{F}$ be a collection of subsets of X. $\mathcal{F}$ is said to be a σ-algebra on X if it satisfies the following assertions.

i. $\emptyset \in \mathcal{F}$.
ii. For any element $A \in \mathcal{F}$, its complement $A^c \in \mathcal{F}$.
iii. For any countable collection $(A_n)_n \subset \mathcal{F}$, the union $\bigcup\limits_n A_n \in \mathcal{F}$.

For any set X there are always two natural examples of σ-algebras on it.

- The one composed of $\{\emptyset, X\}$ known as the coarse σ-algebra.
- The one composed of all subsets of X (the power set of X) denoted $\mathcal{P}(X)$.
- Consequently, any σ-algebra lies between the two types above.

As a consequence of Definition 2.1, the σ-algebra may be seen otherwise.

Proposition 2.1. *Let X be a set, and $\mathcal{F}$ be a σ-algebra on X. The following assertions are true.*

(1) *The whole set X is an element of $\mathcal{F}$, $(X \in \mathcal{F})$.*
(2) *$\mathcal{F}$ is invariant under finite unions.*
(3) *$\mathcal{F}$ is invariant under finite intersections.*
(4) *$\mathcal{F}$ is invariant under countable unions.*
(5) *$\mathcal{F}$ is invariant under the difference of sets.*

Proof. 1. We know, from Assertion (i) Definition 2.1 above, that $\emptyset \in \mathcal{F}$. Hence, by Assertion (ii) in Definition 2.1, its complement $\emptyset^c \in \mathcal{F}$. As $\emptyset^c = X$, we get $X \in \mathcal{F}$.
2. Let $A_0, A_1, \ldots, A_n$, $(n \in \mathbb{N}$ fixed$)$ be a finite collection of elements of $\mathcal{F}$, and consider the countable collection $(B_i)_{i \in \mathbb{N}}$, such that,

$$B_i = A_i, \text{ for } 1 \leq i \leq n, \text{ and } B_i = \emptyset, \text{ for } i \geq n+1.$$

It is straightforward that $(B_i)_{i \in \mathbb{N}} \subset \mathcal{F}$. Hence, Assertion (iii) in Definition 2.1 yields that the union $B_i \in \mathcal{F}$. Now, observe that
$$\small i \in \mathbb{N}$$

$$\bigcup_{i \in \mathbb{N}} B_i = A_0 \cup A_1 \cup \cdots \cup A_n.$$

It results that

$$A_0 \cup A_1 \cup \cdots \cup A_n \in \mathcal{F}.$$

3. Let $A_0, A_1, \ldots, A_n$, $(n \in \mathbb{N}$ fixed$)$ be a finite collection of elements of $\mathcal{F}$, and consider the collection $B_0, B_1, \ldots, B_n$, such that

$$B_i = A_i^c, \text{ for } 1 \leq i \leq n.$$

By Assertion (ii) in Definition 2.1, we conclude that $B_0, B_1, \ldots, B_n$ are elements of $\mathcal{F}$. Assertion 2 above yields that their union

$$B_0 \cup B_1 \cup \cdots \cup B_n \in \mathcal{F}.$$

Now, using again Assertion (ii) in Definition 2.1, we obtain

$$\left(B_0 \cup B_1 \cup \cdots \cup B_n\right)^c \in \mathcal{F}.$$

Observe next that

$$\left(B_0 \cup B_1 \cup \cdots \cup B_n\right)^c = A_0 \cap A_1 \cap \cdots \cap A_n.$$

We thus get

$$A_0 \cap A_1 \cap \cdots \cap A_n \in \mathcal{F}.$$

4. Let $A_0, A_1, \ldots, A_n, \ldots$ be countable collection of elements $A_n \in \mathcal{F}$, and consider the countable collection $(B_n)_{n \in \mathbb{N}}$, such that, $B_n = A_n^c$, $\forall n$. We thus obtain from Assertion (ii), Definition 2.1, that $(B_n)_{n\in\mathbb{N}}$ is a countable collection of elements of $\mathcal{F}$. Hence, Assertion (iii) in Definition 2.1 yields that the union $\underset{n\in\mathbb{N}}{\cup} B_n \in \mathcal{F}$. Now, observe that

$$\underset{n\in\mathbb{N}}{\cup} B_n = \left(\underset{n\in\mathbb{N}}{\cap} A_n\right)^c.$$

It results, from Assertion (ii) in Definition 2.1 again, that $\underset{n\in\mathbb{N}}{\cap} A_n \in \mathcal{F}$.

5. This is an easy consequence of the previous assertions. Indeed, let $A, B \in \mathcal{F}$. Then, $A, B^c \in \mathcal{F}$. It follows, from Assertion 3 above, that $A \cap B^c \in \mathcal{F}$, which is equivalent to $A \setminus B \in \mathcal{F}$. $\qquad\square$

2.2   Some topological concepts

In this subsection, we aim to recall some basic concepts from topology, which will be used later.

Definition 2.2. A topology on a set X is a subset $\mathcal{T}$ of $\mathcal{P}(X)$ satisfying

   i. $\emptyset$, and X are elements of $\mathcal{T}$.
  ii. $\mathcal{T}$ is invariant under arbitrary unions.
 iii. $\mathcal{T}$ is invariant under finite intersections.

The pair $(X, \mathcal{T})$ is said to be a topological space. The elements of $\mathcal{T}$ are called open sets in X. A subset of X is said to be closed if its complementary is open.

As for the case of σ-algebras, for any set X, there are always two natural examples of topology on it.

- The topology $\mathcal{T} = \{\emptyset, X\}$ known as the trivial or the coarse topology.
- The topology $\mathcal{T} = \mathcal{P}(X)$ known as the discrete topology.

Example

(1) $X = \{1, 2, 3\}$. $\mathcal{T} = \{\emptyset, \{1\}, \{2\}, \{1, 2\}, X\}$ is a topology.
(2) $X = \{1, 2, 3\}$, and $\mathcal{T} = \{\emptyset, \{1\}, \{2\}, \{1, 2\}, X\}$. $\mathcal{T}$ is not a topology.
(3) $X = \mathbb{N}$, and $\mathcal{T} = \{A \subset \mathbb{N}, \ A \text{ is finite}\} \cup \{N\}$. $\mathcal{T}$ is not a topology, because (for example) the union of all finite sets, not containing zero, is infinite but is not in $\mathcal{T}$.

One of the important cases of topological spaces are those associated to metrics known as metric spaces.

Definition 2.3. We call metric on a non-empty set X, any mapping $d :$ $X \times X \longrightarrow [0, \infty[$, satisfying

i. $d(x, y) = 0 \iff x = y, \forall x, y \in X$.
ii. $d(x, y) = d(y, x), \forall x, y \in X$.
iii. $d(x, y) \leq d(x, z) + d(z, y), \forall x, y, z \in X$.

The couple (X, d) is called a metric space.

Definition 2.4. Let $(X, \mathcal{T})$ be a topological space. We call a Borel σ-algebra the smallest σ-algebra containing the topology $\mathcal{T}$. The pair $(X, \mathcal{B})$ is called a measurable space.

The following concepts are known in measure theory. Let $(X, \mathcal{B})$ be a measurable space, and $\mu : \mathcal{B} \to [0 + \infty]$ be a positive measure on X.

- If $\mu(X) < +\infty$, μ is said to be finite or bounded measure.
- $\mu(X)$ is called the total mass of μ.
- If $\mu(X) = 1$, μ is called a probability measure.

2.3 Outer measures

The concept of outer measures is a refinement of the notion of measures, in the sense that, it excludes some characteristics. We have precisely the following definition.

Definition 2.5. An outer measure on a set X is any map

$$\mu \colon \mathcal{P}(X) \to [0, +\infty],$$

satisfying the following assertions.

- $\mu(\emptyset) = 0$.
- $\mu(A) \leq \mu(B)$, for all $A, B \subset X$, such that, $A \subset B$.
- $\mu(\bigcup_n A_n) \leq \sum_n \mu(A_n)$, for any sequence $(A_n)_n$ in $\mathcal{P}(X)$.

Definition 2.6. Let μ be an outer measure on X. A subset $A \subset X$ is said to be μ-measurable if

$$\mu(E) = \mu(E \cap A) + \mu(E \cap A^c), \forall E \subset X.$$

Theorem 2.1. *Let μ be an outer measure on X. The collection $\mathcal{B}$ of all μ-measurable of X is a σ-algebra on X. Moreover, the restriction $\mu_{\mathcal{B}}$ of μ on $\mathcal{B}$ is a measure.*

Proof. i. We shall show that

$$\mu(E) = \mu(E \cap X) + \mu(E \cap X^c), \quad \forall E \subset X,$$

and

$$\mu(E) = \mu(E \cap \emptyset) + \mu(E \cap \emptyset^c), \quad \forall E \subset X,$$

which is true as

$$E \cap X = E \cap \emptyset^c = E \quad \text{and} \quad E \cap \emptyset = E \cap X = \emptyset.$$

ii. For all $A \in \mathcal{B}$, we have

$$\mu(E) = \mu(E \cap A) + \mu(E \cap A^c), \quad \forall E \subset X,$$

which is equivalent to

$$\mu(E) = \mu(E \cap A^c) + \mu(E \cap (A^c)^c), \quad \forall E \subset X.$$

Consequently, $A^c \in \mathcal{B}$.

iii. Let $(A_n)_{n \geq 0}$ be a sequence of elements of $\mathcal{B}$. Denote

$$A = \cup_n A_n \text{ and } Y_n = \cup_{k=0}^n A_k.$$

For each $n \geq 0$, the set Y_n is μ-measurable, as finite union of μ-measurable. Moreover, the sequence $(Y_n)_n$ is increasing to A. So, for any subset $E \subset X$, we have for any subset $E \subset X$,

$$\mu(E) = \mu(E \cap Y_n) + \mu(E \cap Y_n^c)$$
$$\geq \mu(E \cap Y_n) + \mu(E \cap A^c).$$

Observe now that the sequence $(Y_n)_n$ is increasing to A. It results that $(\mu(E \cap Y_n))_n$ is increasing to $\mu(E \cap A)$. This completes the proof. $\quad \square$

2.4 Regular outer measures

Definition 2.7. An outer measure μ on X is called regular if $\forall E \subset X$, there exists a μ-measurable set A containing E, and satisfying $\mu(E) = \mu(A)$.

Lemma 2.1. *Let μ be an outer regular measure on X, and $(E_n)_n$ be an increasing sequence of elements of $\mathcal{P}(X)$, then*

$$\mu(\lim_{n \to +\infty} E_n) = \lim_{n \to +\infty} \mu(E_n).$$

Proof. As the sequence $(E_n)_n$ is increasing, it holds that

$$\mu(\lim_{n \to +\infty} E_n) \geq \lim_{n \to +\infty} \mu(E_n).$$

Now, denote $\mathcal{B}$ the σ-algebra of μ-measurable subsets in X. μ being regular, hence, $\forall n \geq 0$, there exists $A_n \supset E_n$, $A_n \subset \mathcal{B}$, such that, $\mu(A_n) = \mu(E_n)$. We have

$$\lim_{n \to \infty} E_n = \liminf_n E_n.$$

Consequently,

$$\mu(\lim_{n \to \infty} E_n) = \mu(\liminf_n E_n) \leq \mu(\liminf_n(A_n)).$$

This yields, using Fatou's Lemma, that

$$\begin{aligned}
\mu(\lim_{n \to \infty} E_n) &\leq \liminf_n \mu(A_n) \\
&= \liminf_{n \to \infty} \mu(E_n) \\
&= \lim_{n \to \infty} \mu(E_n).
\end{aligned}$$

$\square$

2.5 Metric outer measures

Let (X, d) be a metric space.

Definition 2.8. An outer measure μ on X is called metric iff

$$\mu(E \cup F) = \mu(E) + \mu(F), \ \forall E, F \subset X \text{ satisfying } d(E, F) > 0.$$

Proposition 2.2. *Let μ be a outer measure on (X, d), $(A_n)_n \subset X$ an increasing sequence, and*

$$A = \cup_n A_n = \lim_{n \to \infty} A_n.$$

Assume further that

$$d(A_n, A \backslash A_{n+1}) > 0, \ \forall n.$$

Then,

$$\mu(A) = \lim_{n \to \infty} \mu(A_n).$$

Proof. Denote for $k \in \mathbb{N}$,

$$C_k = A_{k+1} \setminus A_k$$

the annulus between A_{k+1}, and A_k. We immediately get

$$d(C_j, C_p) > 0, \ \forall p \geq j + 2,$$

and

$$A = A_n \cup \left(\bigcup_{k=n+1}^{\infty} C_k \right), \ \forall n.$$

Consequently,

$$\mu(A) \leq \mu(A_n) + \sum_{n+1}^{\infty} \mu(A_{k+1} \setminus A_k)$$
$$= \mu(A_n) + R_n,$$

where we denoted for $n \in \mathbb{N}$,

$$R_n = \sum_{n+1}^{\infty} \mu(A_{k+1} \setminus A_k).$$

We shall show that $R_n \longrightarrow 0$ as $n \longrightarrow \infty$. It suffices to prove that the series $\sum_{k \geq 1} \mu(C_{2k})$, and $\sum_{k \geq 1} \mu(C_{2k+1})$ are convergent. Indeed,

$$\sum_{k=1}^{n} \mu(C_{2k+1}) = \mu \left(\bigcup_{k=1}^{n} C_{2k+1} \right)$$
$$\leq \mu(A_{2n+2})$$
$$< \mu(A).$$

Similarly, we do for $\sum_{k \geq 1} \mu(C_{2k})$. As a result,

$$\mu(A) \leq \lim_{n \to \infty} \mu(A_n).$$

$\square$

Theorem 2.2. *Let μ be an outer metric measure on (X, d). Then, every Borel subset of X is μ-measurable.*

Proof. Let B be a Borel subset of X, and denote for $n \geq 1$,

$$B_n = \left\{ x \in E \setminus B \text{ such that } d(x, B) \geq \frac{1}{n} \right\}.$$

It is straightforward that B is a closed, and $E \subset X$. Moreover, $(B_n)_n$ is an increasing sequence. It satisfies also

$$\bigcup_n B_n = E \setminus B,$$

and

$$d(B_n, (E \setminus B) \setminus B_{n+1}) > 0, \ \forall n.$$

It follows, from the Proposition 2.2 above, that

$$\mu(E) \geq \mu(B_n \cup (E \cap B))$$
$$= \mu(B_n) + \mu(E \cap B).$$

Letting $n \longrightarrow \infty$, we obtain

$$\mu(E) \geq \mu(E \setminus B) + \mu(E \cap B).$$

$\square$

2.6 Lebesgue measure on $\mathbb{R}^d$

We designate by $P = (a_1, b_1) \times \cdots \times (a_d, b_d)$, $(a_i < b_i$ real numbers), a cube of order d in $\mathbb{R}^d$, and denote

$$V(P) = \prod_{i=1}^{d} (b_i - a_i)$$

its volume. The Lebesgue measure on $\mathbb{R}^d$ is defined for $E \subset \mathbb{R}^d$ by

$$\mathfrak{m}(E) = \inf \sum V(P_j)/(P_j)_j,$$

where the lower bound is taken on the set of all cubes $(P_j)_j$ covering E $(E \subset \cup_j P_j)$. It is straightforward that

(1) $\mathfrak{m}$ is an outer metric measure on $\mathbb{R}^d$ called Lebesgue outer measure.
(2) $\mathfrak{m}(P) = V(P)$ for all cube P in $\mathbb{R}^d$.
(3) Borel sets are obviously Lebesgue measurable.
(4) $\mathfrak{m}$ is a regular outer measure on $\mathcal{B}_{\mathbb{R}^d}$.

Definition 2.9. The restriction of $\mathfrak{m}$ on the σ-algebra of $\mathfrak{m}$-measurable subsets in $\mathbb{R}^d$ is called the Lebesgue measure in dimension n.

2.7   Convergence of measures on metric spaces

Let (X, d) be a metric space, and $\mathcal{B}$ the Borel σ-algebra on X. Let also μ be a measure on $\mathcal{B}$. Recall that μ is said to be finite if $\mu(X) < \infty$. In particular, μ is said to be a probability measure on X if $\mu(X) = 1$. We recall finally that an element $E \in \mathcal{B}$ is said to be regular relatively to the finite measure μ if it satisfies the assumption

$$\mu(E) = \sup\{\mu(K), \ K \text{ is closed}, \ K \subset E\}$$
$$= \inf\{\mu(O), \ O \text{ is open}, \ O \supset E\}.$$

The measure μ is said to be regular if any element $E \in \mathcal{B}$ is regular. We immediately have a preliminary result on regular measures.

Lemma 2.2. *Let μ be a regular measure on a measurable space $(X, \mathcal{B})$. A set $E \in \mathcal{B}$ is μ-regular if, and only if, $\forall \epsilon > 0$, there exists an open set G, and a closed set F satisfying, $F \subset E \subset G$, and $\mu(G \setminus F) < \epsilon$.*

Proof. $\Longrightarrow$) $E \in \mathcal{B}$ is μ-regular yields that $\forall \epsilon > 0$, there exists a closed set $F_\epsilon \subset E$, such that

$$\mu(E) - \frac{\epsilon}{2} < \mu(F_\epsilon) \leq \mu(E),$$

and

$$\mu(E) \leq \mu(G_\epsilon) < \mu(E) + \frac{\epsilon}{2}.$$

We immediately obtain

$$\mu(E) - \frac{\epsilon}{2} < \mu(F_\epsilon) \leq \mu(E) \leq \mu(G_\epsilon) < \mu(E) + \frac{\epsilon}{2}.$$

As a result,

$$\mu(G_\epsilon \setminus F_\epsilon) = \mu(G_\epsilon) - \mu(F_\epsilon) < \frac{\epsilon}{2} + \frac{\epsilon}{2} = \epsilon$$

$\Longleftarrow$) We observe that $E \setminus F \subset G \setminus F$. Hence,

$$\mu(E \setminus F) \leq \mu(G \setminus F) < \epsilon,$$

which yields that

$$\mu(F) \leq \mu(E) < \mu(F) + \epsilon,$$

or equivalently

$$\mu(E) - \epsilon < \mu(F) \leq \mu(E).$$

Hence,

$$\mu(E) = \sup_{F \subseteq E} \mu(F).$$

Now, observe that similarly that $G\backslash E \subset G\backslash F$. Consequently

$$\mu(G\backslash E) < \epsilon,$$

which yields that

$$\mu(E) \leq \mu(G) < \mu(E) + \epsilon.$$

As a result,

$$\mu(E) = \inf_{G \supset E} \mu(G).$$

$\square$

As a result of this lemma, we obtain the following result.

Proposition 2.3. *Every finite measure μ on $\mathcal{B}$ is regular.*

Proof. Denote $\mathcal{T} = \{E \in \mathcal{B}, E \text{ is } \mu\text{-regular}\}$. We shall show that $\mathcal{T} = \mathcal{B}$. The inclusion $\mathcal{T} \subset \mathcal{B}$ is obvious. Furthermore, $\emptyset, X \in \mathcal{T}$. The collection $\mathcal{T}$ is also invariant by passage to the complement. We will show now that it is invariant under countable unions. So, let $(E_n)_n \subset \mathcal{T}$, denote $E = \bigcup_n E_n$, and let $\epsilon > 0$. It holds immediately that $\forall n$, there exists an open set $O_{n,\epsilon}$, and a closed set $F_{n,\epsilon}$ such that

$$F_{n,\epsilon} \subset E_n \subset O_{n,\epsilon} \text{ and } \mu(O_{n,\epsilon} \backslash F_{n,\epsilon}) < \frac{\epsilon}{3^n}.$$

Let next $O_\epsilon = \bigcup_n O_{n,\epsilon}$. It consists of an open set that contains E. Let also $F = \bigcup_n F_{n,\epsilon}$. As μ is finite, there exists $n_0 \in \mathbb{N}$ such that,

$$\mu(F \backslash \bigcup_{n=1}^{n_0} F_{n,\epsilon}) < \frac{\epsilon}{2}.$$

Consider the set $F_\epsilon = \bigcup_{n=1}^{n_0} F_{n,\epsilon}$. It is obviously a closed set, contained in E. Moreover, we have

$$\mu(O_\epsilon \backslash F_\epsilon) \leq \mu(O_\epsilon \backslash F) + \mu(F \backslash F_\epsilon) \leq \epsilon.$$

Consequently, $\mathcal{T}$ is invariant under countable unions. We thus conclude that $\mathcal{T}$ is a σ-algebra on X. Observe next that for any closed set $F \subset X$, there exists a sequence of open sets $(O_n)_n$ which decreases to F. Hence, there exists $n_1 \in \mathbb{N}$, such that,

$$\mu(O_{n_1}) - \mu(F) < \epsilon, \ \forall \epsilon > 0.$$

Then, F is regular. It result that $\mathcal{T}$ is a σ-algebra, which contains the closed sets. Consequently, $\mathcal{T} \supset \mathcal{B}$. As a result, $\mathcal{T} = \mathcal{B}$.

Definition 2.10. Let $\mathcal{C}_0$ be the set of bounded continuous functions on X, and $(\mu_n)_n$ be a sequence of finite measures on $\mathcal{B}$. We say that $(\mu_n)_n$ converges weakly to a measure μ, and we write $\mu_n \Longrightarrow \mu$, iff

$$\int_X g d\mu_n \longrightarrow \int_X g d\mu, \quad \forall g \in \mathcal{C}_0, \text{ as } n \longrightarrow \infty.$$

Definition 2.11. A set E is said to be a continuity set for μ if $\mu(\partial E) = 0$.

We immediately have the following theorem.

Theorem 2.3. *Let $(\mu_n)_n$ be a sequence of finite measures on $\mathcal{B}$, and μ be a finite measure on $\mathcal{B}$ also. The following assertions are equivalent.*

(1) $\mu_n \Longrightarrow \mu$.
(2) $\mu(C) \geq \overline{\lim}_{n \to \infty} \mu_n(C)$, *for all closed set C in X.*
(3) $\mu(O) \leq \underline{\lim}_{n \to \infty} \mu_n(O)$, *for all open set O in X.*
(4) $\lim_{n \to \infty} \mu_n(E) = \mu(E)$, *for all continuity set E of μ.*

Proof. (**i** $\Rightarrow$ **ii**) Let C be a closed subset in X, and for $p \in \mathbb{N}$, denote

$$S_p = \left\{ x \in X, \ d(x, C) < \frac{1}{2^p} \right\}.$$

Hence, C, and S_p^c are disjoint closed subsets in X. Consequently, there exists a function $g_p \in \mathcal{C}_0$, such that, $0 \leq g_p \leq 1$, $g_p \equiv 1$ on C, and $g_p \equiv 0$ on S_p^c. Now, as $(S_p)_p$ decreases to its limit C, we obtain

$$\overline{\lim_{n \to \infty}} \mu_n(C) = \overline{\lim_{n \to \infty}} \int_X \chi_c d\mu_n$$

$$\leq \overline{\lim_{n \to \infty}} \int_X g_p d\mu_n$$

$$= \int_X g_p d\mu.$$

Next, as $g_p \in \mathcal{C}_0$, it follows, $\forall p \in \mathbb{N}$, that

$$\overline{\lim_{n \to \infty}} \mu_n(C) \leq \int_X g_p d\mu$$

$$= \int_{S_p} g_p d\mu$$

$$\leq \mu(S_p).$$

Letting $p \to +\infty$, we get

$$\varlimsup_{n \to \infty} \mu_n(C) \leq \mu(C).$$

($\mathbf{ii} \Rightarrow \mathbf{iii}$) It is easy by considering the complementary.

($\mathbf{iii} \Rightarrow \mathbf{iv}$) Let E be such that $\mu(\partial E) = 0$. So,

$$\mu(E) = \mu(\overline{E}) = \mu(\overset{\circ}{E}).$$

As $\overline{E}$ is closed, we obtain

$$\varlimsup_{n \to \infty} \mu_n(E) \leq \varlimsup_{n \to \infty} \mu_n(\overline{E}) \leq \mu(\overline{E}) = \mu(E).$$

Similarly, as $\overset{\circ}{E}$ is open, we get

$$\mu(E) = \mu(\overset{\circ}{E}) \leq \varliminf_{n \to \infty} \mu_n(\overset{\circ}{E}) \leq \varliminf_{n \to \infty} \mu_n(E).$$

($\mathbf{iv} \Rightarrow \mathbf{i}$) Fix $\epsilon > 0$, let $g \in \mathcal{C}_0$, and μ^g be defined on the Borel σ-algebra $\mathcal{B}_{\mathbb{R}}$ on $\mathbb{R}$ by

$$\mu^g(A) = \mu\{x \in X / g(x) \in A\}, \ \forall A \in \mathcal{B}_{\mathbb{R}}.$$

It is easy to see that μ^g is a finite measure on $\mathcal{B}_{\mathbb{R}}$. Moreover, as g is bounded, there exists an interval $[a, b] \subset \mathbb{R}$, such that, $\mu^g \equiv 0$ on $[a, b]^c$. Consider next a subdivision $t_0 = a < t_1 < \cdots < t_k = b$, such that $t_{j+1} - t_j < \epsilon$, $\mu^g\{t_j\} = 0$, $\forall j = 1, 2, \ldots, k$, and let

$$E_j = \{x \in X, \ t_{j-1} < g(x) \leq t_j\}.$$

The E_j's are disjoint Borel sets. Moreover, $X = \bigcup_j E_j$. Now, observe that

$$\partial E_j = \{g(x) = t_{j-1}\} \bigcup \{g(x) = t_j\},$$

we deduce that $(\partial E_j) = 0$. Consequently, $\forall j = 1, 2, \ldots, k$ we obtain

$$\mu_n(E_j) \to \mu(E_j) \text{ as } n \to \infty.$$

Denote next $h = \sum_{j=1}^k t_j \chi_{E_j}$. We get

$$\sup_{x \in X} |g(x) - h(x)| < \epsilon.$$

Consequently, for $n \in \mathbb{N}$, we obtain

$$\left| \int_X g d\mu_n - \int_X g d\mu \right| \leq \int_X |g - h| d\mu_n$$
$$+ \left| \int_X h d\mu_n - \int_X h d\mu \right|$$
$$+ \int_X |g - h| d\mu$$
$$\leq \epsilon \mu_n(X) + \epsilon \mu(X)$$
$$+ \sum_{j=1}^k |t_j| |\mu_n(E_j) - \mu(E_j)|.$$

This completes the proof. $\qquad\qquad\qquad\qquad\qquad\qquad\qquad\qquad \square$

Corollary 2.1. *Let μ, ν be finite measures on X, such that,*

$$\int_X g d\mu = \int_X g d\nu$$

for any continuous function $g \in \mathcal{C}_0$. Then, $\mu \equiv \nu$.

Proof. It follows from Theorem 2.3. Indeed, consider the sequence $(\mu_n)_n$, such that, $\mu_n = \nu$. We immediately observe that $\mu_n \Rightarrow \mu$. Hence, for all closed set F in X, we have $\nu(F) \leq \mu(F)$. This means that $\mu \equiv \nu$ on the closed sets in X. Next, as μ, and ν are regular, it follows that $\mu \equiv \nu$ on $\mathcal{B}$. $\qquad\square$

Corollary 2.2. *Let $(\mu)_n$ be a sequence of finite measures on $\mathcal{B}$, μ, and ν be finite measures on $\mathcal{B}$ also, such that, $\mu_n \Rightarrow \mu$, and $\mu_n \Rightarrow \nu$. Then, $\mu \equiv \nu$.*

Proof. It follows from Corollary 2.1 immediately. Indeed, $\forall g \in \mathcal{C}_0$, we get

$$\int_X g \, d\mu_n \underset{n \to +\infty}{\to} \int_X g \, d\mu,$$

and

$$\int_X g \, d\mu_n \underset{n \to +\infty}{\to} \int_X g \, d\nu.$$

Hence,

$$\int_X g \, d\mu = \int_X g \, d\nu, \forall g.$$

Therefore $\mu \equiv \nu$. $\qquad\square$

Definition 2.12. Let $\mathcal{M} = \{\mu_t, \ t \in \mathcal{T}\}$ be a collection of finite measures on $\mathcal{B}$.

(1) $\mathcal{M}$ is said to be relatively compact, if any sequence $(\mu_n)_n$ in $\mathcal{M}$ has a sub-sequence convergent weakly to a finite measure on $\mathcal{B}$.
(2) $\mathcal{M}$ is said to be tight, if for all $\epsilon > 0$, there exists a compact $K_\epsilon \subset X$, such that, $\sup_{t \in \mathcal{T}} \mu_t(X \setminus K_\epsilon) < \epsilon$.

Example Consider $X = \mathbb{R}$, and $\mathcal{F} = \{f_t, \ t \in \mathcal{T}\}$ a collection of bounded functions on $\mathbb{R}$, which are decreasing, and right continuous,, and $\mathcal{M}$ the set of corresponding finite measures.

The following theorem due to Prokorov relates the concept of relatively compact measures, and the tight ones.

Theorem 2.4. *Prokorov Theorem.* *Let (X, d) be a complete separable metric space. Let $\mathcal{B}_X$ be the Borel σ-algebra on X, and $\mathcal{M}$ be a collection of probability measures on $\mathcal{B}_X$. Then, $\mathcal{M}$ is relatively compact if, and only if $\mathcal{M}$ is tight.*

Proof. $\Longrightarrow$) Assume that $\mathcal{M}$ is relatively compact,, and let $\epsilon > 0$. We shall show that, there exists a compact $K_\epsilon \subset X$, such that

$$\mu(K_\epsilon) > 1 - \epsilon, \ \forall \mu \in \mathcal{M}.$$

Let $k \in \mathbb{N}$. As X is separable, there exists $\{x_n\}_n \subset X$ dense in X. Denote so

$$S_{n,k} = \{x \in X, \ d(x_n, x) < \frac{1}{k}\}.$$

We observe that $\bigcup_n S_{n,k} = X$. Denote next for $n \in \mathbb{N}$,

$$U_{n,k} = \bigcup_{j=1}^{n} S_{j,k}.$$

It consists of a sequence of open sets increasing to X. Assume thus that there exists ϵ_0, such that, $\forall n$, there exists a measure μ_n satisfying

$$\mu_n(U_{n,k}) \leq 1 - \frac{\epsilon_0}{2^k}.$$

As $\{\mu_n\} \subset \mathcal{M}$, there exists a sub-sequence $\mu_{(\varphi(n))}$ convergent weakly to $\mu \in \mathcal{M}$. We immediately get

$$\begin{aligned} \mu(U_{s,k}) &\leq \varliminf_{n\to\infty} \mu_{\varphi(n)}(U_{s,k}) \\ &\leq \varliminf_{n\to\infty} \mu_{\varphi(n)}(U_{\varphi(n),k}) \\ &\leq 1 - \frac{\epsilon_0}{2^k}. \end{aligned}$$

As a consequence,

$$\mu(U_{s,k}) \leq 1 - \frac{\epsilon_0}{2^k}, \ \forall s.$$

Letting $s \to \infty$, we get

$$\mu(X) \leq 1 - \frac{\epsilon_0}{2^k},$$

which contradicts the fact that $\mu(X) = 1$. Therefore, there exists $n_k \leq 1$, such that,

$$\mu(U_{n_k,k}) > 1 - \frac{\epsilon_0}{2^k}, \ \forall \mu \in \mathcal{M}.$$

So, denote

$$K_\epsilon = \bigcap_{k \geq 1} \bigcup_{j=1}^{n_k} S_{j,k}.$$

It is now straightforward that K_ϵ is compact, and

$$\mu(K_\epsilon) > 1 - \epsilon, \ \forall \mu \in \mathcal{M}.$$

$\Longleftarrow$). Assume that $\mathcal{M}$ is tight. So, $\forall n \geq 0$, there exists a compact K_n in X, such that,

$$\mu(K_n) > 1 - \frac{1}{2^n}.$$

Now, as X is separable, there exists a countable set of open balls $U = \{S_p, p \geq 0\}$, which constitutes a neighborhoods' basis of X. Denote so

$$\mathcal{F} = \left\{ \bigcup_{\text{finite}} \overline{S} \bigcap K_n \ S \in \mathcal{U} \right\}.$$

It is easy to check that $\mathcal{F}$ is countable, invariant under finite union, and its elements are compact. Hence, for any sequence $(\mu_n)_n \subset \mathcal{M}$, there exists a sub-sequence $(\mu_{n_k})_n \subset (\mu_n)_n$, for which, $\lim\limits_{k \to \infty} \mu_{n_k}(F)$ exists, and finite for all $F \in \mathcal{F}$. Denote next

$$\nu(F) = \lim\limits_{k \to +\infty} \mu_{n_k}(F).$$

Define now the measure μ on $\mathcal{B}$ by

$$\mu(O) = \sup_{F \subset O, \ F \in \mathcal{F}} \nu(F), \tag{2.1}$$

for all open set O in X. Observing that

$$\nu(F) = \lim\limits_{k \to +\infty} \mu_{n_k}(F) \leq \underline{\lim}_k \mu_{n_k}(O),$$

we deduce that

$$\mu(O) \leq \underline{\lim}_k \mu_{n_k}(O).$$

As a consequence, Theorem 2.3 yields that $\mu_{n_k} \Longrightarrow \mu$. So, it suffices to prove that ν defines well a probability measure μ on X. To do this, consider the mapping μ^* defined on $\mathcal{P}(X)$ by

$$\mu^*(E) = \inf_{O \supset E, \ O \ \text{open}} \mu(O), \ \ \forall E \in \mathfrak{P}(X).$$

As $\nu(\phi) = 0$, we immediately get $\mu^*(\phi) = 0$. Next, for $E_1 \subset E_2$ in $\mathcal{P}(X)$, let O be an open set containing E_2. We have

$$\mu^*(E_1) \leq \mu(O),$$

which yields that

$$\mu^*(E_1) \leq \mu^*(E_2).$$

Let now $(E_n)_n$ be a countable set in $\mathcal{P}(X)$, and denote $E = \bigcup E_n$. Observe next that, for all $\epsilon > 0$, and all $n \in \mathbb{N}$, there exists open sets O_n, such that

$$\mu(O_n) \leq \mu^*(E_n) + \frac{\epsilon}{2^n},$$

which, with the fact that $E \subset \bigcup_n O_n$, yields that

$$\mu^*(E) \leq \mu(\bigcup_n O_n).$$

It suffices so to show that

$$\mu(\bigcup_n O_n) \leq \sum_n \mu(O_n). \tag{2.2}$$

To do it, let $F \in \mathcal{F}$ be, such that, $F \subset \bigcup_n O_n$. As F is compact, there exists $n_0 \in \mathbb{N}$, such that, $F \subset \bigcup_{n=1}^{n_0} O_n$, which, by the next, yields that,

$$\nu(F) \leq \mu(\bigcup_{n=1}^{n_0} O_n).$$

Hence, it suffices to show that (2.2) holds for a set of two elements O_1, and O_2 in $\mathcal{F}$, such that, $F \subset O_1 \cup O_2$. Denote thus

$$C_1 = \left\{ x \in F, \ d(x, O_1^c) \geq d(x, O_2^c) \right\},$$

and

$$C_2 = \left\{ x \in F, \ d(x, O_2^c) \geq d(x, O_1^c) \right\}.$$

It is easy to see that C_1, and C_2 are compact, $C_1 \subset O_1$, and $C_2 \subset O_2$. On the other hand, there exists $x_{1,j}..., x_{N_j,j} \in C_1$, such that

$$C_j \subset \bigcup_{s=1}^{N_j} S_{x_s,j} \subset \bigcup_{s=1}^{N_j} \overline{S}_{x_s,j} \subset \theta_j, \quad 1 \leq j \leq 2.$$

So, as $F \in \mathcal{F}$, there exists compact sets K_{n_j}, satisfying $F \subset K_{n_j}$, which implies that

$$C_j \subset \bigcup_{s=1}^{N_j} \overline{S}_{x_s,j} \cap K_{n_j} \subset O_1.$$

Denote next

$$F_j = \bigcup_{s=1}^{N_j} \overline{S}_{x_s,j} \cap K_{n_j}.$$

Hence, $F_j \in \mathcal{F}$. Moreover, there exists $F_1, F_2 \in \mathcal{F}$, such that,

$$C_1 \subset F_1 \subset O_1, \quad C_2 \subset F_2 \subset O_2,$$

and that

$$F \subset C_1 \bigcup C_2 \subset F_1 \bigcup F_2 \subset O_1 \bigcup O_2.$$

As a result,

$$\nu(F) \leq \nu(F_1) + \nu(F_2) \leq \mu(O_1) + \mu(O_2),$$

which yields that

$$\mu(O_1 \bigcup O_2) \leq \mu(O_1) + \mu(O_2).$$

Consequently, whenever $F \subset \bigcup_n O_n$, we get

$$\nu(F) \leq \mu(\bigcup_{n=1}^{n_0} O_n) \leq \sum_{j=1}^{n_0} \mu(O_j) \leq \sum_{j \geq 1} \mu(O_j).$$

Hence,

$$\mu(\bigcup_n O_n) \leq \sum_n \mu(O_n).$$

As a result,

$$\mu^\star(\bigcup_n E_n) \leq \mu(\bigcup_n \theta_n) \leq \sum_n \mu(\theta_n) \leq \mu^\star(\bigcup_n E_n) + \epsilon.$$

We thus proved that $\mu^\star$ is an outer measure. It remains to show that the σ-algebra

$$\mathcal{B}^\star = \{E \in \mathcal{B},\ E \text{ is } \mu^\star\text{-measurable}\} = \mathcal{B}.$$

So, let $A \in \mathcal{P}(X)$. We have

$$\mu^\star(E) \leq \mu^\star(E \cap A) + \mu^\star(E \cap A^c).$$

Let next $\epsilon > 0$, and $O \supset E$, O being an open set. There exists $F_0 \in \mathcal{F}$, such that,

$$F_0 \subset O \cap A^c, \text{ and } \mu(O \cap A^c) < \nu(F_0) + \frac{\epsilon}{2}.$$

Let $F_1 \in \mathcal{F}$ with $F_1 \subset O \cap F_0^c$, and

$$\mu(O \cap F_0^c) < \nu(F_1) + \frac{\epsilon}{2}.$$

We immediately get

$$F_0 \cap F_1 = \emptyset, \text{ and } F_0 \cup F_1 \subset O.$$

Hence,

$$\mu(\theta) \geq \nu(F_0 \cup F_1)$$
$$= \nu(F_0) + \nu(F_1)$$
$$\geq \mu(O \cap A^c) - \frac{\epsilon}{2} + \mu(O \cap F_0^c) - \frac{\epsilon}{2}.$$

Equivalently,

$$\mu(O) \geq \mu(O \cap A^c) + \mu(O \cap A) - \epsilon$$
$$\geq \mu^\star(O)$$
$$\geq \mu^\star(O \cap A^c) + \mu(O \cap A) - \epsilon.$$

This is true for all open set $O \supset E$. As a result,

$$\mu(O) \geq \mu(O \cap A^c) + \mu(O \cap A) - \epsilon.$$

It holds, from all the previous developments, that $\mathcal{B}^\star$ is a σ-algebra that contains the closed sets in X, and thus it contains the Borel σ-algebra $\mathcal{B}$. Moreover, we observe that $\mu = \mu^\star_{|\mathcal{B}}$, and that

$$1 \geq \mu(X) \geq \sup_n \mu(K_n) \geq \sup_n (1 - \frac{1}{2^n}) = 1.$$

Hence, $\mu(X) = 1$, and consequently, μ is a probability measure on X. $\quad\square$

2.8   Exercises for Chapter 2

Exercise 1.
Let $(A_n)_n$ be a sequence of elements of a σ-algebra $\mathcal{F}$ on X. Denote

$$\overline{\lim_n} A_n = \bigcup_{n \geq 0} (\bigcap_{k \geq n} A_k),$$

and

$$\underline{\lim_n} A_n = \bigcap_{n \geq 0} (\bigcup_{k \geq n} A_k).$$

Show that

(1) $\overline{\lim_n} A_n \subset \underline{\lim_n} A_n$.

(2) $\overline{\lim_n} A_n \in \mathcal{F}$.

(3) $\underline{\lim_n} A_n \in \mathcal{F}$.

Exercise 2.
Let $(X, \mathcal{B})$ be a measurable space, and $\mu : \mathcal{B} \to [0 + \infty]$ be a positive measure on X. Prove the following assertions.

(1) Whenever $A, B \in \mathcal{B}$ such that $A \subseteq B$, then $\mu(A) \leq \mu(B)$.

(2) μ is σ-additive,, in the sense that, for any sequence $(A_n)_{n \in \mathbb{N}} \subset B$, we have

$$\mu(\bigcup_{n \in \mathbb{N}} A_n) \leq \sum_{n \in \mathbb{N}} \mu(A_n).$$

(3) For any increasing sequence $(A_n)_{n \in \mathbb{N}} \subset B$ we have

$$\mu(\lim_{n \to \infty} A_n) = \lim_{n \to \infty} \mu(A_n).$$

(4) Whenever $(A_n)_{n \in \mathbb{N}} \subset B$ is a decreasing sequence, for which, there exist an element A_{n_0}, such that, $\mu(A_{n_0}) < \infty$, we have

$$\mu(\lim_{n \to \infty} A_n) = \lim_{n \to \infty} \mu(A_n).$$

(5) For any sequence $(A_n)_{n \geq 0} \subset B$ we have

$$\mu(\liminf_{n \to \infty} A_n) \leq \liminf_{n \to \infty} \mu(A_n).$$

Exercise 3.
Prove the assertions 1 to 4 in subsection 2.6.

Exercise 4.
Let μ be a metric outer measure on (X, d). Let A_{nn} be an increasing sequence of subsets of X with $A = \lim_n A_n$, and assume that

$$d(A_n, A \setminus A_{n+1}) > 0, \, \forall n.$$

Show that $\mu(A) = \lim_n \mu(A_n)$.

Exercise 5.
Let $\mathbb{Q} = (q_1, q_2, ...)$ be an enumeration of the rational numbers. For $A \subset \mathbb{R}$, let

$$\mu(A) = \sum_{q_i \in A} 2^{-i}.$$

(1) Show that μ is a measure on $\mathbb{R}$.
(2) Show that all subsets of $\mathbb{R}$ are μ-measurable.
(3) Show that $\mu(\mathbb{R}|\mathbb{Q}) = 0$.

Exercise 6.
Let μ be a measure on $\mathbb{R}^n$, such that, for all $x \in \mathbb{R}^n$, there is a ball $B(x, r)$ with $\mu(B(x, r)) < \infty$. Show that μ is locally finite.

Exercise 7.

Let $f : [0,1] \to \mathbb{R}^+$ be continuous,, and define μ on $[0,1]$ by

$$\mu(A) = \int_A f(x)dx.$$

Show that μ is equivalent to the Lebesgue measure.

Exercise 8.

Show that, if $\mu_k \to \mu$, and if A is a bounded set with $\mu(\partial A) = 0$, where ∂A is the boundary of A, then $\mu_k(A) \to \mu(A)$.

Exercise 9.

Let $r > 0$, and consider in $\mathbb{R}^2$ the rectangles of the form

$$C_{m,n}(r) = (nr, (n+1)r) \times (mr, (m+1)r),$$

where n, m are integers. Let also μ be a finite measure on $\mathbb{R}^2$, and $\alpha \geq 0$. We write

$$N_r(\alpha) = \sharp\{C_{m,n}(r); \ \mu(C_{m,n}(r)) \geq r^\alpha\},$$

and define the functions

$$f_c(\alpha) \equiv \lim_{\epsilon \to 0} \lim_{r \to 0} \frac{\log^+(N_r(\alpha + \epsilon) - N_r(\alpha - \epsilon))}{-\log r},$$

$$\underline{f}_c(\alpha) \equiv \lim_{\epsilon \to 0} \liminf_{r \to 0} \frac{\log^+(N_r(\alpha + \epsilon) - N_r(\alpha - \epsilon))}{-\log r},$$

and

$$\overline{f}_c(\alpha) \equiv \lim_{\epsilon \to 0} \limsup_{r \to 0} \frac{\log^+(N_r(\alpha + \epsilon) - N_r(\alpha - \epsilon))}{-\log r},$$

for $\alpha \geq 0$. Show that

$$f_H(\alpha) \leq \underline{f}_c(\alpha) \leq \overline{f}_c(\alpha), \ \forall \alpha \geq 0.$$

Exercise 10.

A $-$ Let λ be the Lebesgue measure on $[-1,1]$, and $g = \chi_{[0,1]}$. Let next (f_n) be the sequence of functions defined by

$$\begin{cases} f_n(x) = g(x) & \text{for } n \text{ even}, \\ f_n(x) = g(-x) & \text{for } n \text{ odd}. \end{cases}$$

Show that

$$\int \left(\liminf_{n \to \infty} f_n \right) < \liminf_{n \to \infty} \int f_n.$$

B – Let $(X, \mathcal{F}, \mu)$ be a measurable space. For a real-valued measurable function f on X, denote

$$\Phi_f(t) = \mu(\{f > t\}), \ t \geq 0.$$

(1) Show that if f is a step function, then

$$\int_{\mathbb{R}_+} \Phi_f(t) dt = \int_X f d\mu.$$

(2) Show that for every real-valued measurable function f on X, and for all $p > 0$, we have

$$\int_X |f|^p d\mu = p \int_{\mathbb{R}_+} \mu(\{|f| > t\}) t^{p-1} dt.$$

Chapter 3

Martingales with Discrete Time

The notion of martingales constitutes one of the basic concepts applied in fractal analysis. We propose in this chapter to provide a detailed study of such a notion in order to conserve the self-containing aspect of the book. Readers may refer for more details, and for general studies of martingales concept and applications to [Benaych-Georges (2009); Berglund (2014); Berglund and Gentz (2006); Billingsley (1965); Budzinski (2019); Falconer (1985, 1990); Gervais (2009); Rogers (1970)].

3.1 Generalities

Let $(\Omega, \mathcal{A}, P)$ be a probability space. We will denote by $L(\Omega, \mathcal{A}, P)$ the vector space of all real $\mathcal{A}$-measurable functions defined on Ω. Such functions are known as random variables. For $p \in [1, +\infty[$, we denote

$$L^p(\Omega, \mathcal{A}, P) = \left\{ f \text{ which are } \mathcal{A} - \text{measurable, and } \int_\Omega |f|^p dP < +\infty \right\}.$$

We next define the p-norm on $L^p(\Omega, \mathcal{A}, P)$ by

$$\|f\|_p = \left(\int_\Omega |f|^p dP \right)^{\frac{1}{p}}.$$

Let $\mathcal{B}$ be a complete sub-σ-field in $\mathcal{A}$. As previously, we denote

$$L(\mathcal{B}) = \{ f \text{ which are } \mathcal{B} - \text{measurable} \},$$

the vector space of all $\mathcal{B}$-measurable functions on Ω. For $1 \le p < \infty$, we introduce similarly the $L^p(\mathcal{B})$ vector space by

$$L^p(\mathcal{B}) = \{ f \in L(\mathcal{B}); \|f\|_p < +\infty \}.$$

We just recall here that there is no essential difference with the functions that are complex-valued, or taking their values in the higher dimensional

Euclidean space $\mathbb{R}^d$, for $d \geq 2$. In these cases, the measurability of the whole function is equivalent to the measurability of all its components, and the absolute value will be seen as the module in the complex case or the norm in $\mathbb{R}^d$. As a consequence, all the functions used will be real-valued, except when necessary will recalled. We have the following properties of functional spaces above.

Proposition 3.1.

(1) $L(\mathcal{B})$ *(resp $L^p(\mathcal{B}), 1 \leq p \leq \infty$) is the sublinear space of $L(\Omega, \mathcal{A}, P)$ (resp $L^p(\Omega, \mathcal{A}, P)$) and is reticulated. It is invariant with respect to monotone limits.*

(2) *For $1 \leq p < \infty$, $L^p(\mathcal{B})$ is closed.*

Proof. Let $f, g \in L(\mathcal{B})$. We claim that $\max(f, g) \in L(\mathcal{B})$ and $\min(f, g) \in L(\mathcal{B})$. Indeed, for $c \in \mathbb{R}$, we have

$$(\max(f, g))^{-1}(-\infty, c) = f^{-1}(] - \infty, c[) \bigcup (g^{-1}(] - \infty, c[) \in \mathcal{B}.$$

Similarly,

$$(\min(f, g))^{-1}(-\infty, c) = f^{-1} - \infty, c[) \bigcap (g^{-1}(] - \infty, c[) \in \mathcal{B}.$$

Hence, $L(\mathcal{B})$ is reticulated. Next, for all $f, g \in L^p(\mathcal{B})$, we have

$$\| \max(f, g) \|_p = \| \frac{f + g + |f - g|}{2} \|_p \leq \|f\|_p + \|g\|_p < +\infty,$$

and

$$\| \min(f, g) \|_p = \| \frac{f + g - |f - g|}{2} \|_p \leq \|f\|_p + \|g\|_p < +\infty.$$

Hence, $\max(f, g)$ and $\min(f, g) \in L^p(\mathcal{B})$. Let next $(f_n)_n$ be Increasing in $L(\mathcal{B})$. We shall show that its limit $f \in L(\mathcal{B})$. (Respectively, for $(f_n)_n$ in $L^p(\mathcal{B})$, we show that its limit $f \in L^p(\mathcal{B})$). We claim that, if M is a sublinear space in $L^p(\mathcal{B})$, satisfying the conditions of Proposition 3.1, there exists a sub-σ-field $\mathcal{B}$ of A, such that, $M = L(\mathcal{B})$ (respectively, $M = L^p(\mathcal{B})$). So, assume, for the second part of the Proposition 3.1, that $1 \leq p < +\infty$, so that, $L^p(\mathcal{B})$ will be closed in $L^p(\Omega, \mathcal{A}, P)$. $\qquad \square$

Remark 3.1. The space $\mathcal{C}([0, 1], \mathbb{R}) \subseteq L^\infty([0, 1], \mathbb{R})$, within the Lebesgue measure, is a counter example to the second part of Proposition 3.1, for $p = \infty$.

Indeed, the sequence $(f_n)_n$ defined by

$$f_n(x) = \begin{cases} 0 & , \ 0 \le x \le \frac{1}{2} - \frac{1}{2n}, \\ 2nx + 1 - n & , \ \le \frac{1}{2} - \frac{1}{2n} \le x \le \frac{1}{2}, \\ 1 & , \ \frac{1}{2} \le x \le 1, \end{cases}$$

lies in $\mathcal{C}([0,1],\mathbb{R})$. Its limit being the function

$$f(x) = \begin{cases} 0 \ , \ 0 \le x < \frac{1}{2}, \\ 1 \ , \ \frac{1}{2} \le x \le 1, \end{cases}$$

is not in $\mathcal{C}([0,1],\mathbb{R})$, even though it is in $L^\infty([0,1],\mathbb{R})$.

Corollary 3.1. *Let $1 \le p < \infty$, and U be a positive contractive linear operator on $L^p(\Omega, \mathcal{A}, P)$, such that $U1 = 1$. Then, there exists a complete sub-σ-field $\mathcal{J}$, satisfying $L^p(\mathcal{J}) = \{f, Uf = f\}$.*

Proof. It is straightforward that the set $V = \{f, Uf = f\}$ is a linear subspace of L^p. Moreover, it is closed and contains the constant function 1. Now, as U is a positive operator, we immediately deduce that

$$U(f^+) \ge Uf, \quad \text{and} \quad U(f^+) \ge 0, \ \forall f \in L^p.$$

This yields that

$$U(f^+) \ge (Uf)^+, \ \forall f \in L^p.$$

Consequently,

$$U(f^+) \ge f^+ \ge 0, \ \forall f \in V.$$

As a result, due to the fact that U is a contraction, we get

$$U(f^+) = f^+, \ \forall f \in V.$$

Observing now that

$$Sup(f,g) = f + (g - f)^+,$$

it results that V is reticulate. $\square$

Corollary 3.2. *Let $\{\mathcal{B}_n, \ n \in \mathbb{N}\}$ be an Increasing sequence of sub-σ-fields of $\mathcal{A}$, and $\mathcal{B}_\infty = \mathcal{T}(\bigcup_n \mathcal{B}_n)$. Then,*

$$L^p(\mathcal{B}_\infty) = \overline{\bigcup_{n \in \mathbb{N}} L^p(\mathcal{B}_n)}, \ \forall p, \ 1 \le p < \infty.$$

Proof. Observe that $\overline{\bigcup_{n\in\mathbb{N}} L^p(\mathcal{B}_n)}$ is reticulated, closed, and it contains the constant function 1. Hence, by Proposition 3.1, there exists a sub-σ-field $\mathcal{B} \subseteq \mathcal{A}$, such that,

$$L^p(\mathcal{B}) = \overline{\bigcup_{n\in\mathbb{N}} L^p(\mathcal{B}_n)}.$$

We will show next that $\mathcal{B} = \mathcal{B}_\infty$. Indeed, $\{\mathcal{B}_n,\ n \in \mathbb{N}\}$ being increasing, we thus deduce that $\bigcup_{n\in\mathbb{N}} L^p(\mathcal{B}_n)$ is a linear subspace of L^p, which is reticulate, and contains the constant function 1. Now, observe that the functions $f \to af$, $(f,g) \to f+g$ and $(f,g) \to f \vee g$, $(f,g) \to f \wedge g$ are continuous. We thus deduce that the closure $\overline{\bigcup_{n\in\mathbb{N}} L^p(\mathcal{B}_n)}$ is also a linear subspace of L^p, closed, reticulate, and contains the constant function 1. It results from the previous corollary that there exists a complete sub-σ-field $\mathcal{B}$ of $\mathcal{A}$, such that

$$\overline{\bigcup_{n\in\mathbb{N}} L^p(\mathcal{B}_n)} = L^p(\mathcal{B}).$$

Observe next that

$$L^p(\mathcal{B}_n) \subset L^p(\mathcal{B}), \quad \forall n \in \mathbb{N}.$$

As a result, due to the fact that $\mathcal{B}$ is complete, we get

$$\mathcal{B}_n \subset \tilde{\mathcal{B}}_n \subset \mathcal{B}.$$

Consequently,

$$\mathcal{B}_\infty \subset \mathcal{B},$$

which yields that

$$L^p(\mathcal{B}_\infty) \subset L^p(\mathcal{B}).$$

On the other hand, we have

$$\mathcal{B}_n \subset \mathcal{B}_\infty, \quad \forall n \in \mathbb{N}.$$

Therefore,

$$L^p(\mathcal{B}_n) = L^p(\mathcal{B}_\infty)(n \in \mathbb{N}),$$

which means that

$$L^p(\mathcal{B}_n) \subset L^p(\mathcal{B}_\infty), \quad (n \in \mathbb{N}).$$

Hence,

$$L^p(\mathcal{B}) = \overline{\bigcup_{n\in\mathbb{N}} L^p(\mathcal{B}_n)} \subset L^p(\mathcal{B}_\infty).$$

We get finally,

$$L^p(\mathcal{B}) = L^p(\mathcal{B}_\infty),$$

and that $\mathcal{B} = \tilde{\mathcal{B}}_\infty$. $\qquad\qquad\qquad\qquad\qquad\qquad\qquad\qquad\qquad\qquad\square$

Corollary 3.3. *Let $(\mathcal{B}_n)_{n\in\mathbb{N}}$ be a decreasing sequence of sub-σ-fields in $\mathcal{A}$. Then, for $1 \le p \le \infty$, we have*

$$\bigcap_{n\in\mathbb{N}} L^p(\mathcal{B}_n) = L^p\left(\bigcap_{n\in\mathbb{N}} \mathcal{B}_n\right).$$

Proof. See Exercise 3.6. □

3.2 Conditional expectation

Let $\mathcal{B}$ be a sub-σ-field in $\mathcal{A}$, and consider on $L^2(\Omega, \mathcal{A}, P)$ the scalar product defined by

$$< f, g > = \int_\Omega fg dP, \quad \forall f, \ g \in L^2(\Omega, \mathcal{A}, P).$$

We have immediately the main result proved in the theory of Hilbert spaces, stating that for every $f \in L^2(\Omega, \mathcal{A}, P)$, there exists a unique element $\tilde{f} \in L^2(\mathcal{B})$, such that,

$$< \tilde{f}, g > = < f, g >, \quad \forall g \in L^2(\mathcal{B}).$$

$\tilde{f}$ is called the projection of f onto $L^2(\mathcal{B})$, and it satisfies the condition of smallest distance from $L^2(\mathcal{B})$ to f, where the distance between two elements $f, g \in L^2(\Omega, \mathcal{A}, P)$ is defined by $\|f - g\|_2$. In all that follows, we denote $\tilde{f}$ by $E^{\mathcal{B}}(f)$.

Definition 3.1. The function $E^{\mathcal{B}} : f \longmapsto E^{\mathcal{B}}(f)$ is called conditional expectation relatively to $\mathcal{B}$.

We immediately have the following results.

Proposition 3.2.

(1) $\forall f \in L^2(\Omega, \mathcal{A}, P), \ \forall B \in \mathcal{B}, \ \int_B E^{\mathcal{B}}(f)dP = \int_B f dP.$

(2) $\forall f \in L^2(\Omega, \mathcal{A}, P), \ \forall h \in L^\infty(\mathcal{B}), \ E^{\mathcal{B}}(hf) = hE^{\mathcal{B}}(f).$

Proof. 1. We have for $g = X_B$,

$$< f, g > = \int_B f dP$$

$$< E^{\mathcal{B}}(f), g > = \int_B E^{\mathcal{B}}(f)dP.$$

2.

$$< hf, g > = < hE^{\mathcal{B}}(f), g >, \quad \forall g.$$

$$\int_\Omega hfgdP = \int_\Omega fhgdP$$
$$= < E^{\mathcal{B}}(f), hg >$$
$$= \int_\Omega fE^{\mathcal{B}}(f)hgdP$$
$$= < hE^{\mathcal{B}}(f), g > .$$

We now prove that $hE^{\mathcal{B}}(f)$ satisfies effectively the properties of characterizing the conditional expectation $E^{\mathcal{B}}(hf)$ of hf. We know that $hE^{\mathcal{B}}(f) \in L^2(\mathcal{B})$, and is a product of $\mathcal{B}$ – measurable functions is L^∞ and L^2, respectively. On the other hand, for any element $g \in L^2(\mathcal{B})$, the function $hg \in L^2(\mathcal{B})$. As a consequence,

$$\int_\Omega fE^{\mathcal{B}}(f) \, hg \, dP = \int_\Omega fhgdP,$$

which means that $E^{\mathcal{B}}(f)h - fh$ is orthogonal to $L^2(\mathcal{B})$. $\qquad\square$

Proposition 3.3. *Let U be an orthogonal projector on $L^2(\mathcal{B})$. Then U is a conditional expectation if and only if U is positive, and all constant functions remain invariant within the action of U.*

Proof. It is easy to observe that $E^{\mathcal{B}}(1) = 1$. Hence, to prove that $E^{\mathcal{B}}(f) \geq 0$ for $f \geq 0$, we have to use the function

$$1_{(E^{\mathcal{B}}(f)<0)} \in L^2(\mathcal{B}).$$

We obtain

$$0 \leq \int f1_{(E^{\mathcal{B}}(f)<0)}dP = \int_{(E^{\mathcal{B}}(f)<0)} E^{\mathcal{B}}(f)dP \leq 0.$$

The two integrals are zero. Henceforth, $\{E^{\mathcal{B}}(f) < 0\}$ is of zero-measure, which yields that $E^{\mathcal{B}}(f) \geq 0$ in L^2.

Conversely, assume that U is a positive orthogonal projector on L^2, for which $U1 = 1$. Then, the set (vector space) $\{f : Uf = f\}$ takes the form $L^2(\mathcal{B})$ fort some complete sub-σ-algebra $\mathcal{B}$ in $\mathcal{A}$. As U is an orthogonal projector on $\{f : Uf = f\}$, we get $U = E^{\mathcal{B}}$. $\qquad\square$

Lemma 3.1. *Let $\bar{L}_+(\mathcal{A}) = \{f : \Omega \to \bar{\mathbb{R}}_+, f \text{ is } \mathcal{A}\text{-measurable}\}$, and $\mathcal{B}$ a sub σ-field in $\mathcal{A}$. Denote $\bar{L}_+(\mathcal{B}) = \{f \in \bar{L}_+(\mathcal{A}), f \text{ is } \mathcal{B}\text{-measurable}\}$. Then, $\forall f \in \bar{L}_+(\mathcal{B})$, there exists a unique $\tilde{f} \in \bar{L}_+(\mathcal{B})$, such that,*

(1) $\displaystyle\int_\Omega \tilde{f}gdP = \int_\Omega fgdP, \forall g \in \bar{L}_+(\mathcal{B}).$

(2) $\displaystyle\int_B \tilde{f}dP = \int_B fdP, \forall B \in \mathcal{B}.$

Proof. Assume firstly that $f \in L^2_+$. By the definition of E^B on L^2, $\tilde{f}$ exists, and $\tilde{f} = E^B(f)$. Let $g \in \bar{L}_+(\mathcal{B})$, and denote $g_n = \inf(g, n)$, $n \in \mathbb{N}$. Hence, $\forall n$, $g_n \in L^2(\mathcal{B})$, and $(g_n)_n \nearrow g$ in $L^2(\mathcal{B})$. Therefore,

$$\int_\Omega E^B(f) g_n dP = \int_\Omega f g_n dP.$$

The theorem of monotone convergence implies that

$$\int_\Omega E^B(f) g dP = \int_\Omega f g dP, \ \forall f \in \bar{L}_+(\mathcal{B}), \ \forall g \in \bar{L}_+(\mathcal{B}).$$

Let now $f \in \bar{L}_+$ and $g \in \bar{L}_+(\mathcal{B})$ and put for $n \in \mathbb{N}, f_n = \inf(f, n)$. So, $(f_n) \nearrow f$ in L^2_+. E^B is in fact a positive linear operator on L^2. It is then an increasing operator on L^2. Consequently, $(E^B(f_n))_n$ is an increasing sequence in $L^2(\mathcal{B})$. Denote, $\hat{f}$ its pointwise limit in $L^2(\mathcal{B})$. Then the monotone convergence theorem implies that

$$\int_\Omega \hat{f} g dP = \int_\Omega f g dP.$$

We thus take $\hat{f} = \tilde{f}$. Then assertion **1** holds.

To finish the proof of the lemma, we notice that $\mathbf{1} \Rightarrow \mathbf{2}$. □

3.3 Convergence and regularity of martingales

In this section, we will prove the fundamental theorem on the convergence of martingales and upper-martingales. Besides, we introduce the notion of regularity and regular stopping time.

Definition 3.2. Let $(X_n)_{n \in \mathbb{N}}$ be an adapted sequence of real random variables, such that, X_n is integrable, $\forall n \in \mathbb{N}$.

- The sequence $(X_n)_n$ is said to be a sub-martingale if

$$\forall n, \ X_n \leq E^{\mathcal{B}_n}(X_{n+1}) \text{ almost surely.}$$

- The sequence $(X_n)_n$ is said to be integrable martingale if

$$X_n = E^{\mathcal{B}_n}(X_{n+1}), \ \forall n.$$

Let X be a real random variables. We denote

$$X^+ = \sup(X, 0) \quad \text{and} \quad X^- = \sup(-X, 0).$$

We have immediately, $X = X^+ - X^-$. Remark that $E(X^+) < \infty$ allows us to define the conditional expectation $E^{\mathcal{B}}(X)$ as

$$E^{\mathcal{B}}(X) = E^{\mathcal{B}}(X^+) - E^{\mathcal{B}}(X^-).$$

Let $(X_n)_n$ be an integrable sub-martingale, such that $\sup_n E(X_n^+) < \infty$. Hence, $(X_n^+)_n$ is a positive integrable sub-martingale. Indeed, observe in one hand, that

$$E(X_n) = E(X_n^+) - E(X_n^-).$$

On the other hand, we also have

$$X_p \le E^{\mathcal{B}_p}(X_{p+1}^+), \ \forall p.$$

Consequently,

$$X_p^+ \le E^{\mathcal{B}_p}(X_{p+1}^+), \ \forall p.$$

Thus, $(X_n^+)_n$ is a positive integrable sub-martingale. Consider now the sequence $(E^{\mathcal{B}_n}(X_p^+))_{p \ge n}$. We have, $\forall p \ge n$,

$$E^{\mathcal{B}_n}(X_{p+1}^+) = E^{\mathcal{B}_p}\left(E^{\mathcal{B}_n}(X_{p+1}^+)\right)$$
$$= E^{\mathcal{B}_n}\left(E^{\mathcal{B}_p}(X_{p+1}^+)\right)$$
$$\ge E^{\mathcal{B}_n}(X_p^+).$$

This means that the sequence $(E^{\mathcal{B}_n}(X_p^+))_{p \ge n}$ is Increasing with respect to p. Denote, for $n \in \mathbb{N}$,

$$M_n = \sup_{p \ge n} E^{\mathcal{B}_n}(X_p^+).$$

We get

$$E^{\mathcal{B}_n}(M_{n+1}) = E^{\mathcal{B}_n}(\lim_p \uparrow E^{\mathcal{B}_{n+1}}(X_p^+))$$
$$= \lim_p \uparrow E^{\mathcal{B}_n}(E^{\mathcal{B}_{n+1}}(X_p^+))$$
$$= \lim_p \uparrow E^{\mathcal{B}_n}(X_p^+).$$

Consequently,

$$E^{\mathcal{B}_n}(M_{n+1}) = M_n, \ \forall n.$$

On the other hand,

$$E(M_n) = E(\lim_p E^{\mathcal{B}_n}(X_p^+))$$
$$= \lim_p E(E^{\mathcal{B}_n}(X_p^+))$$
$$= \lim_p E(X_p^+)$$
$$\le \sup_p E(X_p^+).$$

As $E(X_p^+) < \infty$, we immediately observe that, $E(M_n) < \infty$. This shows that $(M_n)_n$ is a positive, and integrable martingale. Denote next,

$$Y_n = M_n - X_n, \ n \in \mathbb{N}.$$

Then, $(Y_n)_n$ is a positive, and integrable upper-martingale. By applying the convergence criterion of upper-martingales, we may assume that the almost surely limits

$$M_\infty = \lim_{n \to \infty} M_n \quad \text{and} \quad Y_\infty = \lim_{n \to +\infty} Y_n$$

exist. More precisely, as $\sup_n E(X_n^+) < \infty$, we get

$$E(M_\infty) \leq \underline{\lim}_n E(M_n) < \infty.$$

Similarly, we conclude also that

$$E(Y_\infty) \leq E(M_1) < \infty.$$

This shows that $(M_n)_n$ and $(y_n)_n$ are positive, integrable martingales, and upper-martingales respectively. We thus conclude that the sequence $(X_n)_n$ converges almost surely to $X_\infty = M_\infty - Y_\infty$.

3.4   Regularity of integrable martingales

We know that in general an almost sure convergence doesn't imply the convergence in the space L^1. Thus, our main aim in this part is to search for some conditions to establish the equivalence between the two types of convergence.

Lemma 3.2. *Let X be a real random variable in L^1. Then, the family $\left(E^{\mathcal{B}}(X)\right)$, where $\mathcal{B}$ is a sub-σ-field in $\mathcal{A}$, is an equi-integrable family, in the sense that*

$$\sup_{\mathcal{B}} \int_{|E^{\mathcal{B}}(X)| \geq a} |E^{\mathcal{B}}(X)| dP \downarrow 0, \ as \ a \uparrow \infty.$$

Proof. We have

$$E^{\mathcal{B}}(X) \leq E^{\mathcal{B}}(|X|), \ \forall \mathcal{B}.$$

Hence,

$$
\int_{|E^{\mathcal{B}}(X)|\geq a} |E^{\mathcal{B}}(X)|dP \leq \int_{|E^{\mathcal{B}}(X)|\geq a} E^{\mathcal{B}}(|X|)dP
$$
$$
\leq \int_{|E^{\mathcal{B}}(X)|\geq a} |X|dP
$$
$$
= \int_{\{|E^{\mathcal{B}}(X)|\geq a\}\cap\{|X|\geq\sqrt{a}\}} |X|dP
$$
$$
+ \int_{\{|E^{\mathcal{B}}(X)|\geq a\}\cap\{|X|<\sqrt{a}\}} |X|dP
$$
$$
\leq \int_{|X|\geq\sqrt{a}} |X|dP + \sqrt{a}P(E^{\mathcal{B}}(|X|)\geq a)
$$
$$
\leq \int_{|X|\geq\sqrt{a}} |X|dP + \frac{\sqrt{a}}{a}E(|X|).
$$

Now, as $X \in L^1$, we get

$$
\int_{|X|\geq\sqrt{a}} |X|dP \to 0, \text{ as } a \uparrow \infty.
$$

The second term goes to 0 as $a \to \infty$ obviously. Hence, the proof of the lemma. $\qquad\square$

Lemma 3.3. *Let $(X_n)_n$ be an equi-integrable sequence of a real-valued random variable converging almost surely to a limit X_∞. Then, $(X_n)_n$ converges to X_∞ in the space L^1.*

Proof. Let $a \in \mathbb{R}_+$, and consider the function

$$
f(x) = x\chi_{[-a,a]}(x) + a\chi_{]a,+\infty)}(x) - a\chi_{]-\infty,-a[}.
$$

It is straightforward that

$$
|x - f_a(x)| \leq |x|, \; \forall x \in \mathbb{R}.
$$

We claim that $(X_n)_n$ is a Cauchy's sequence in L^1. Indeed, for $m, n \in \mathbb{N}$, we have

$$
\|X_m - X_n\|_1 \leq \|X_m - f_a(X_m)\|_1
$$
$$
+ \|f_a(X_m) - f_a(X_n)\|_1
$$
$$
+ \|f_a(X_n) - X_n\|.
$$

By the dominated convergence theorem, we conclude that

$$
\|f_a(X_m) - f_a(X_n)\|_1 \to 0 \text{ as } m, n \to \infty.
$$

On the other hand, $\forall m \in \mathbb{N}$, we have

$$
\|X_m - f_a(X_m)\|_1 \leq 2\int_{|X_m|\geq a} |X_m|dP \to 0 \text{ as } a \to \infty.
$$

Hence, $(X_n)_n$ satisfies the Cauchy's criterion in L^1, which is a Banach space. So, we conclude that $(X_n)_n$ converges in L^1. $\qquad\square$

Next, before tackling the main purpose of this section, we firstly prove the fundamental Doob's theorem.

Theorem 3.1. *Doob Theorem.* *Let $(X_n)_n$ be an integrable sub-martingale, satisfying $\sup_n E(X_n^+) < \infty$. Then, $(X_n)_n$ converges almost surely to a limit X in L^1.*

Proof. Consider $X = X_\infty$ defined in Section 1, and M_n, Y_n, such that $X_n = M_n - y_n$. □

Proposition 3.4. *Let $(X_n)_n$ be an integrable martingale. The following assertions are equivalent.*

(1) $(X_n)_n$ *converges in L^1.*
(2) $\sup_n E(|X_n|) < +\infty$, *and* $X_n = E^{\mathcal{B}_n}(X_\infty)$, $\forall n$.
(3) *There exists an integrable real-valued random variable X, such that,*

$$X_n = E^{\mathcal{B}_n}(X), \ \forall n.$$

(4) $(X_n)_n$ *satisfies the property of equi-integrability.*

Proof. $1 \Rightarrow 2$) Let $X \in L^1$ be such that,

$$\|X_n - X\|_1 \to 0 \text{ as } n \to \infty.$$

We have, for n large enough,

$$\|X_n\| < 1 + \|X\|_1 \ \forall \, n \geq n_0.$$

Hence,

$$\sup_n E(|X_n|) < \infty.$$

Consequently,

$$X_\infty = \lim_{n \to +\infty} (a - sX_n),$$

exists, and $X_\infty \in L^1$ due to Doob theorem. Therefore,

$$E^{\mathcal{B}_n}(X_p) \to E^{\mathcal{B}_n}(X_\infty),$$

in L^1 as $(p \to +\infty)$. On the other hand, for $p \geq n$, we have

$$E^{\mathcal{B}_n}(X_p) = X_n.$$

So that,

$$X_n = E^{\mathcal{B}_n}(X_\infty), \ \forall n.$$

2 ⇒ 3) We take $X = X_\infty$. By Fatou's Lemma, we get

$$E(|X_\infty|) \leq \underline{\lim} E(|X_n|) \leq \sup_n E(|X|) < +\infty.$$

Hence, ⇒ $X_\infty \in L^1$.

3 ⇒ 4) Lemma 1 with X yields that

$$\sup_n \int_{|X_n| \geq a} |X_n| dP \to 0 \quad \text{if } a \to +\infty.$$

4 ⇒ 1) is a consequence of Lemma 2.   □

Definition 3.3. An integrable martingale $(X_n)_n$ satisfying one of the assertions of Proposition 3.4 above is said to be regular.

3.5   Positive and upper martingales

Let $(\Omega, \mathcal{A}, P)$ be a probability space and $(\mathcal{B}_n)$ be the increasing sequence of sub-σ-algebras of $\mathcal{A}$ where for $n \in \mathbb{N}$, $\mathcal{B}_n$ is given by

$$\mathcal{B}_n = \mathcal{B}(\{Y_m, m \leq n\}),$$

for some given sequence $(Y_m)_m$ of random variables. Denote

$$\mathcal{B}_\infty = \mathcal{T}(\bigcup_n \mathcal{B}_n).$$

Definition 3.4. A sequence $(X_n)_n$ of random variables is said to be adapted to $\mathcal{B}_\infty$ if for all n X_n is $\mathcal{B}_n$-measurable.

We remark immediately that if $(X_n)_n$ is an adapted sequence and $(f_n)_n$ a sequence of measurable functions, the sequence $(\tilde{X}_n = f(X_1, ..., X_n))_n$ is also adapted. In fact, talking about a random variable X or its expectation $E^{\mathcal{B}}(X)$ is considered as equivalence classes. We will say that X or $E^{\mathcal{B}}(X)$ are measurable by replacing the equivalence class with a $\mathcal{B}$-measurable element of this class, and thus, any property of X or $\mathcal{B}(X)$ is considered as almost surely.

3.5.1   *Stopping time*

The concept of stopping time is strongly related to probability theory such as the experience of coin threatening where the player needs to guess and/or estimates the time of winning or losing the game. Mathematically speaking, the stopping time is also described by a random event in a probability $\{\Omega, \mathcal{A}, (\mathcal{B}_n)_n\}$. The $\mathcal{B}_n$'s inform us about the event time-wise. A stopping rule then consists of giving a rule to quit the game. The rule is based on the information that we have in our hands in every moment n. By denoting ν the stopping instant of a game we are able to introduce the following definition.

Definition 3.5. A map $\nu : \Omega \to \bar{\mathbb{N}}$ is said to be a stopping time if the event $\{\nu = n\} \in \mathcal{B}_n$, $\forall n \in \bar{\mathbb{N}}$. We denote $\mathcal{B}_\nu = \{B \in \mathcal{B}_\infty,\ B \cap \{\nu = n\} \in \mathcal{B}_n, \forall n\}$. An event in $\mathcal{B}_\nu$ is said to be prior to ν.

Remark 3.2. $\mathcal{B}_\nu$ is a sub-σ-algebra of $\mathcal{B}_\infty$.

Corollary 3.4. *Let $(X_n)_n$ be a regular martingale. The following assertions hold.*

(1) For all stopping time ν, X_ν is integrable.
(2) $\forall\ \nu_1, \nu_2$ stopping times, we have

$$X_{\nu_1} = E^{\mathcal{B}_{\nu_1}}(X_{\nu_2}).$$

Proof. It is immediate as we have

$$E^{\mathcal{B}_{\nu_1}}(X_{\nu_2}) = E^{\mathcal{B}_{\nu_1}}(E^{\mathcal{B}_{\nu_2}}(X_\infty)) = E^{\mathcal{B}_{\nu_1}}(X_\infty) = X_{\nu_1}.$$

$\square$

Proposition 3.5. *Let $p > 1$, and $(X_n)_n$ be a martingale in L^p, such that, $\sup_n \|X_n\|_p < \infty$. Then, $(X_n)_n$ is regular and $X_n \underset{n}{\to} X_\infty$ in L^p.*

Proof. Denote $C = \sup_n \|X_n\|_p$. We have

$$\sup_n \int_{\{|X_n| \geq a\}} |X_n| dP \leq \frac{(\sup_n \|X_n\|_p)^p}{a^{p-1}} = \frac{C^p}{a^{p-1}} \underset{a \to \infty}{\to} 0 .$$

Therefore, $(X_n)_n$ is equi-integrable. Consequently, it is regular. Besides, we have by Fatou's Lemma

$$\int_{\{|X_n| \geq a\}} |X_\infty|^p dP \leq \varliminf_n \int_{\{|X_n| \geq a\}} |X_n|^p dP < \infty.$$

Hence, X_∞^p.

$\square$

Lemma 3.4.

(1) $\nu : \Omega \to \bar{\mathbb{N}}$ is a stopping time if and only if $\{\nu \leq n\} \in \mathcal{B}_n$, $\forall n \in \mathbb{N}$.
(2) If $\nu : \Omega \to \bar{\mathbb{N}}$ is a stopping time, and $B \in \mathcal{B}_\infty$, then $B \in \mathcal{B}_\nu$ if and only if $B \cap \{\nu \leq n\} \in \mathcal{B}_n$, $\forall n \in \mathbb{N}$.

Proof. The lemma follows from the fact that

$$\{\nu \leq n\} = \bigcap_{m \leq n} \{\nu = m\} \text{ and } \{\nu = n\} = \{\nu \leq n\} \bigcap \{\nu \leq n-1\}^c.$$

$\square$

Proposition 3.6.

(1) Let ν be a stopping time, and $f : \Omega \to \bar{\mathbb{R}}$ be $\mathcal{B}_\infty$ – measurable. Then, f is $\mathcal{B}_\nu$ – measurable if and only if $f_{|\{\nu=n\}}$ is $\mathcal{B}_n$ – measurable, $\forall n$.

(2) Let g be a positive real-valued measurable or integrable function on $(\Omega, \mathcal{A}, P)$. Then, $E^{\mathcal{B}_\nu}(g) = E^{\mathcal{B}_n}(g)$ on $\{\nu = n\}$, $\forall n \in \bar{\mathbb{N}}$.

Proof 1. It results from the definition of $\mathcal{B}_\nu$ for $f = \chi_B$, where $B \in \mathcal{B}_\infty$. The general case follows immediately.

2. It holds that $E^{\mathcal{B}_n}(g) \geq 0$, $\mathcal{B}_\nu$ – measurable. Moreover, if $B \in \mathcal{B}_\nu$, we have

$$\int_B E^{\mathcal{B}_n}(g)dP = \sum_{p \in \bar{\mathbb{N}}} \int_{B \cap \{\nu=p\}} E^{\mathcal{B}_n}(g)dP$$

$$= \sum_{p \in \bar{\mathbb{N}}} \int_{B \cap \{\nu=p\}} g dP$$

$$= \int_B g dP$$

$$= \int_B E^{\mathcal{B}_\nu}(g)dP.$$

In general, for g integrable, we write $g = g^+ - g^-$, where $g*+ = \max(g,0)$ and $g^- = \max(-g,0)$. $\square$

Corollary 3.5. Let $(X_n)_n$ be an adapted sequence of random variables, and ν be a stopping time. Consider $X_\nu : \{\nu < +\infty\} \to \mathbb{R}$, with

$$X_\nu(w) = X_{\nu(w)}(w), \ \forall w.$$

Then, X_ν is $\mathcal{B}_\nu$-measurable, and satisfies

$$X_\nu \equiv X_n \text{ on } \{\nu = n\}, \ \forall n \in \mathbb{N}.$$

Proof.

$$X_\nu^{-1}(\nu \leq p) = \bigcup_{n \leq p} X_\nu^{-1}(\nu = n)$$

$$= \bigcup_{n \leq p} X_n^{-1}(\nu = n)$$

$X_n^{-1}(\nu = n) \in \mathcal{B}_n, \forall n$. Then $X_\nu^{-1}(\nu \leq p) \in \mathcal{B}_p$.

The second part is obvious, as $\forall w \in (\nu = n)$, we have $\nu(w) = n$. So, $X_\nu(w) = X_n(w)$. $\square$

Proposition 3.7. *Let ν and ν' be two stopping times. We have*

(1) $\{\nu < \nu'\}, \{\nu = \nu'\}, \{\nu \leq \nu'\} \in \mathcal{B}_\nu \bigcap \mathcal{B}_{\nu'}$.
(2) $B \in \mathcal{B}_\nu \implies B \bigcap \{\nu \leq \nu'\} \in \mathcal{B}_{\nu'}$.
(3) $\nu \leq \nu' \implies \mathcal{B}_\nu \subset \mathcal{B}_{\nu'}$.

Proof. (1) There holds that

$$(\nu < \nu') = \bigcup_{\nu=n} (n < \nu') \ \in \mathcal{B}_{\nu'},$$

and

$$(\nu < \nu') = \bigcup_{\nu'=n} (\nu < n) \ \in \mathcal{B}_\nu.$$

(2) $B \cap (\nu \leq \nu') \subset (\nu \leq \nu') \ \in \mathcal{B}_{\nu'}$.
(3) $\forall B \in \mathcal{B}_\nu$, we have $B \cap (\nu \leq \nu') \ \in \mathcal{B}_{\nu'}$. On the other hand, $(\nu \leq \nu') = \Omega$. Hence, we set $B \in \mathcal{B}_{\nu'}$.

$\square$

3.5.2 Positive upper martingales

Definition 3.6. An adapted sequence $(X_n)_{n \in \mathbb{N}}$ of positive random variables is said to be a positive upper-martingale if

$$X_n \geq E^{\mathcal{B}_n}(X_{n+1}), \ \forall n \in \mathbb{N}.$$

The sequence is said to be a positive martingale in the case of the equality.

Such definition explains the Decreasing behavior of the sequence $(X_n)_n$ in terms of the conditional expectation. More precisely, we have the following remark.

Remark 3.3. Whenever $(X_n)_n$ is a positive upper martingale, it holds that

$$X_m \geq E^{\mathcal{B}_m}(X_p), \ \forall m \leq p.$$

Lemma 3.5. *Let $(X_n^{(1)})_n$ and $(X_n^{(2)})_n$ be two positive upper martingales, and ν a stopping time, such that, $X_\nu^{(1)} \geq X_\nu^{(2)}$ on $\{\nu < \infty\}$, $n \in \mathbb{N}$. Consider next*

$$X_n : w \longmapsto X_n(w) = \begin{cases} X_n^{(1)}(w) & \text{for } n < \nu(w), \\ X_n^{(2)}(w) & \text{for } n \geq \nu(w). \end{cases}$$

Then, $(X_n)_n$ is a positive upper martingale.

Proof. It suffices to write

$$X_n = 1_{\{\nu > n\}} X_n^{(1)} + 1_{\{\nu \leq n\}} X_n^{(2)}.$$

$\square$

Proposition 3.8. *The Maximal Inequality.* *Let $(X_n)_n$ be a positive upper martingale. Then, $\sup_n X_n$ is a random variable, which is finite almost surely on $\{X_0 < +\infty\}$, and satisfies*

$$\forall a > 0, \ P^{\mathcal{B}_0}\{\sup_n X_n \geq a\} \leq \min(\frac{X_0}{a}, 1).$$

We write $P^{\mathcal{B}}(A)$ to designate $E^{\mathcal{B}}(1_A)$, where A is a subset of Ω.

Proof. Let $\nu_a : \Omega \longrightarrow \bar{\mathbb{N}}$ be such that,

$$\nu_a(w) = \begin{cases} \min\{n, X_n(w) > a\} & \text{if } (\sup_n X_n)(w) > a, \\ \infty & \text{if } (\sup_n X_n)(w) \leq a. \end{cases}$$

It holds, $\forall p \in \mathbb{N}$, that

$$\{\nu_a = p\} = \{w, X_n(w) > a, \forall n \geq p, X_m(w) \leq a, \forall m \leq p-1\}$$
$$= \left(\bigcap_{n \geq p}\{X_n > a\}\right) \bigcap \left(\bigcap_{m \leq p-1}\{X_n \leq a\} \in \mathcal{B}_p\right).$$

For $p = \infty$, we have

$$\{\nu_a = \infty\} = \{w/X_n(w) < a, \forall n\} = \bigcap_n \{X_n \leq a\} \in \mathcal{B}_\infty.$$

Then, ν_a is a stopping time. As on $\{\nu_a < +\infty\}$, we have $X_{\nu_a} > a$, and the constant a defines a positive upper martingale, it holds that $(Y_n)_n$ given by

$$Y_n = X_n \text{ for } n < \nu \quad \text{and} \quad Y_n = a \text{ for } n \geq \nu_a,$$

is a positive upper martingale. We thus have

$$Y_0 \geq E^{\mathcal{B}_0}(Y_n) \Rightarrow X_0 \geq E^{\mathcal{B}_0}(Y_n) \quad \text{and} \quad a > E^{\mathcal{B}_0}(Y_n).$$

Consequently,

$$\min(X_0, a) \geq E^{\mathcal{B}_0}(Y_n) \geq E^{\mathcal{B}_0}(a1_{\nu_a \leq n}).$$

As a result,

$$\min(X_0, a) \geq aE^{\mathcal{B}_0}(1_{\nu_a \leq n}) = aP^{\mathcal{B}_0}(\nu_a \leq n).$$

Letting $n \longrightarrow \infty$, we obtain

$$\min(X_0, a) \geq aP^{\mathcal{B}_0}(\nu_a < \infty) = aP^{\mathcal{B}_0}(\sup_n X_n > a).$$

Hence,

$$\min(\frac{X_0}{a}, 1) \geq P^{\mathcal{B}_0}(\sup_n X_n > a).$$

$\square$

Corollary 3.6. *As a consequence of the previous results, we have the following.*

(1) $P(X_0 < \infty, \sup_n X_n > a) \leq \int_{\{X_0 < \infty\}} \min(\frac{X_0}{a}, 1)dP.$

(2) $P(X_0 < \infty, \sup_n X_n = \infty) = 0.$

(3) *The constant a may be replaced by a positive random variable A.*

Proof. (1) By integrating in Proposition 3.8 on $\{X_0 < \infty\}$, which belongs already to $\mathcal{B}_0$, we obtain

$$P(X_0 < \infty, \sup_n X_n > a) \leq \int_{\{X_0 < \infty\}} \min(\frac{X_0}{a}, 1)dP.$$

(2) Whenever $a \uparrow \infty$, the second term goes to 0 (by using the dominated convergence Theorem). We thus obtain

$$P(X_0 < \infty, \sup_n X_n = \infty) = 0.$$

(3) It suffices to take $y_n = \dfrac{X_n}{A}$ and $a = 1$. $\qquad\square$

Consider next a sequence $(x_n)_n$ in $\mathbb{R}$, and $a < b$ two fixed real numbers. Define the sequence of integers $(\nu_k)_{k \geq 1}$ by

$$\nu_1 = \min(n; n \geq 0, x_n \leq a),$$

for $k \in \mathbb{N}$,

$$\nu_{2k} = \min(n; n \geq \nu_{2k-1}, x_n \geq b),$$

and

$$\nu_{2k+1} = \min(n; n \geq \nu_{2k}, x_n \leq a).$$

Whenever an index ν_k is not defined, (for example if $x_n > a$, $\forall n$), we put $\nu_k = \infty$ as well as all ν_j, $j \geq k$. We designate by $\beta_{a,b}$ the greatest integer p, such that, ν_{2p} is finite . $\beta_{a,b} = \infty$ whenever all ν_k are finite . $\beta_{a,b}$ designates in fact the number of times that the sequence $(x_n)_n$ crosses the interval $[a, b]$. We have the following result.

Proposition 3.9.

(1) $\varliminf_n x_n < a < b < \varlimsup_n \beta_{a,b} \Rightarrow \beta_{a,b} = \infty \Rightarrow \varliminf_n x_n \leq a < b \leq \varlimsup_n x_n.$

(2) *The sequence $(x_n)_n$ is convergent* $\Longleftrightarrow \beta_{a,b} < \infty, \forall a, b \in \mathbb{R}.$

Consider now a sequence $(X_n)_n$ of positive continuous random variables. It holds that the positive real numbers a, b suffice instead of taking them with arbitrary signs in $\mathbb{R}$.

By the previous procedure, for any element $w \in \Omega$, we associated a sequence of indices $(\nu_k(w))_k$, which are associated in turn to the real sequence $(x_n = X_n(w))_n$, and thus, we defined a sequence of random variables $(\nu_k)_k$. We obtain the following characterizations for the sequence $(\nu_k)_k$,

$$\{\nu_{2p} = n\} = \bigcup_{m<n} \{\nu_{2p-1} = n, X_{m+1} < b, ..., X_{n-1} < b, X_n \geq b\},$$

and

$$\{\nu_{2p+1} = n\} = \bigcup_{m<n} \{\nu_{2p} = n, X_{m+1} < b, ..., X_{n-1} < b, X_n \geq b\}.$$

Now, as $\{\beta_{a,b} \geq p\} = \{\nu_{2p} < \infty\}$, we deduce that $\beta_{a,b}$ is also a random variable, and that

$$\{(X_n)_n \text{ is convergent}\} = \bigcap_{a<b} \{\beta_{a,b} < +\infty\}.$$

As a consequence, we obtain the following theorem.

Theorem 3.2. *A sequence of real random variables $(X_n)_n$ converges almost surely if and only if $\beta_{a,b} < \infty$ almost surely, $\forall a, b \in \mathbb{R}$ or $\mathbb{Q}$.*

Proof. For a sequence $(x_n, n\mathbb{N})$ in $\mathbb{R}$, and a couple of real number (a, b), we define the sequence of integers ν_k, $k \in \mathbb{N}$ recursively as follows,

$$\nu_1 = \min(n \geq 0; \ x_n \leq a),$$

$$\nu_2 = \min(n \geq \nu_1; \ x_n \geq b),$$

$$\nu_3 = \min(n \geq \nu_2; \ x_n \leq a), \ ...$$

and so on. Whenever, one of the values ν_k is not well defined, for example, $x_n 1 > a$, for all k, we fix it equal to ∞, and all the value after it. Denote so, $\beta_{a,b}$ the greatest integer p for which $\nu_{2p} < \infty$, and conventionally, $\beta_{a,b} = \infty$ whenever $\nu_k < \infty$, for all k. It holds that

$$\liminf_{n\to\infty} x_n < a < b < \limsup_{n\to\infty} x_n \implies \beta_{a,b} = \infty$$

$$\implies \liminf_{n\to\infty} x_n \leq a < b \leq \limsup_{n\to\infty} x_n.$$

This yields that $(x_n)_n$ has a limit (in $[-\infty, \infty]$) if and only if $\beta_{a,b} < \infty$ for all real (or also rational) numbers $a < b$.

Consider next a sequence $(X_n)_n$ of real-valued random variables. To each event ω, we have a sequence $(\nu_k(\omega))_k$ defined as above. To such a sequence, we correspond the sequence $(X_n(\omega))_n$. Notice that

$$(\nu_{2p} = n) = \bigcup_{m<n} (\nu_{2p-1} = m; \ X_{m+1} < b, \ldots, X_{n-1} < b, X_n \geq b),$$

and similarly for the odd indices. Next, as $(\nu_{2p} < \infty) = (\beta_{a,b} \geq p)$, we get the convergence criterion for the random variable $\beta_{a,b}$, as

$$(X_n \longrightarrow) = \bigcap_{a<b \,\text{in}\, \mathbb{R}} (\beta_{a,b} < \infty) = \bigcap_{a<b \,\text{in}\, \mathbb{Q}} (\beta_{a,b} < \infty).$$

$\square$

Corollary 3.7. *Let $(X_n)_n$ be a positive upper-martingale, and let $\beta_{a,b}$ be the associated numbers of Increasing crossing times. Then, $\forall\, k \geq 1$ integer, and $\forall\, a, b, \ 0 < a < b < \infty$, we have*

$$P^{\mathcal{B}_0}[\beta_{a,b} \geq k] \leq \left(\frac{a}{b}\right)^k \min\left(\frac{X_0}{a}, 1\right).$$

Moreover, $\beta_{a,b}$ are finite almost surely.

Proof. Consider the sequence $(Y_n)_n$ defined as follows. For $k \geq 1$ in $\mathbb{N}$,

$$Y_n = \begin{cases} 1 & \text{for } 0 \leq n < \nu_1, \\ \dfrac{X_n}{a} & \text{for } \nu_1 \leq n < \nu_2, \\ \dfrac{b}{a} & \text{for } \nu_2 \leq n < \nu_3, \\ \dfrac{b}{a}\dfrac{X_n}{a} & \text{for } \nu_3 \leq n < \nu_4, \\ \vdots \\ \left(\dfrac{b}{a}\right)^{k-1}\dfrac{X_n}{a} & \text{for } \nu_{2k-1} \leq n < \nu_{2k}, \\ \left(\dfrac{b}{a}\right)^k & \text{for } \nu_{2k} \leq n. \end{cases}$$

It holds easily that $(Y_n)_n$ is a positive upper martingale, $\forall k$ fixed. Observe next that

$$Y_0 = \min\left(\frac{X_0}{a}, 1\right),$$

and that

$$Y_n \geq \left(\frac{b}{a}\right)^k 1_{\nu_{2k} \leq n}, \quad \forall n.$$

Consequently, as $Y_0 \geq E^{\mathcal{B}_0}(y_n)$, we obtain

$$\left(\frac{b}{a}\right)^k P^{\nu_{2k} \leq n} \leq \min\left(\frac{X_0}{a}, 1\right).$$

Letting $n \to \infty$, this yields that

$$(\frac{b}{a})^k P^{\mathcal{B}_0}(\nu_{2k} < \infty) \leq \min(\frac{X_0}{a}, 1).$$

Letting now $k \to \infty$, it results that

$$(\frac{b}{a})^k P^{\mathcal{B}_0}(\mathcal{B}_{a,b} \geq k) \leq \min(\frac{X_0}{a}, 1),$$

which implies finally that $\mathcal{B}_{a,b} < \infty$ almost surely. $\square$

As a consequence we obtain the following theorem.

Theorem 3.3. *Any upper positive martingale $(X_n)_n$ converges almost surely. Furthermore, the limit $X_\infty = \lim\limits_{n \to \infty} X_n$ satisfies*

$$E^{\mathcal{B}_n}(X_\infty) \leq X_n, \ \forall n \in \mathbb{N}.$$

Proof. From above, it suffices to show that $\mathcal{B}_{a,b} < \infty$, almost surely, for all $0 < a < b < \infty$, which is a consequence of Subin's inequality, which states that, $\forall n > p$,

$$E^{\mathcal{B}_p}(\underline{\lim}_{m \geq n}) \leq E^{\mathcal{B}_p}(X_n) \leq X_p.$$

Letting next $n \to \infty$, we obtain

$$E^{\mathcal{B}_p}(X_\infty) \leq X_p, \ \forall p \in \mathbb{N}.$$

     $\square$

Proposition 3.10. *Let $p \in [1, \infty[$, $Z \in L_+^p$. Then, the sequence $(Z_n = E^{\mathcal{B}_n}(Z))_n$ is a positive martingale which is convergent almost everywhere, and in L^p to the random variable $Z_\infty = E^{\mathcal{B}_\infty}(Z)$.*

Proof. $(Z_n)_n$ is indeed a martingale, and, for all n, we have

$$E^{\mathcal{B}_n}(Z_{n+1}) = E^{\mathcal{B}_n}(E^{\mathcal{B}_{n+1}}(Z)) = E^{\mathcal{B}_n}(Z) = Z_n.$$

Therefore, Theorem 3.3 yields that $(Z_n)_n$ converges almost surely to a limit Z_∞, which is $\mathcal{B}_\infty$-measurable. So, let next $A \in \bigcup_n \mathcal{B}_n$, $m \in \mathcal{N}$, such that, $A \in \mathcal{B}_m$, and let finally $n \geq m$. We get

$$\int_A E^{\mathcal{B}_n}(Z)dP = \int_A ZdP.$$

On the other hand,

$$\int_A E^{\mathcal{B}_n}(Z)dP = \int_A Z_n dP \longrightarrow \int_A Z_\infty dP.$$

Consequently, $\forall A \in \mathcal{B}_\infty$, we get

$$\int_A Z_\infty dP = \int_A Z dP \Rightarrow Z_\infty = E^{\mathcal{B}_\infty}(Z).$$

Now, whenever Z is upper bounded, $0 \leq Z \leq a$, it holds that $Z_\infty \leq a$, and thus, $Z_\infty \in L^p+$, and $\|Z_n - Z\|_p \to 0$ whenever $n \to \infty$. Now, observe that Z may be decomposed as

$$Z = \underbrace{\min(Z, a)}_{X} + \underbrace{\max(Z, a)}_{Y},$$

where $a > 0$ in $\mathbb{R}$. Hence,

$$Z_n = E^{\mathcal{B}_n}(Z) = E^{\mathcal{B}_n}(X) + E^{\mathcal{B}_n}(Y) = X_n + Y_n.$$

Observe next that X and Y are positive in L^p, and satisfy further $X \leq a$. Therefore,

$$\|E^{\mathcal{B}_n}(Z) - E^{\mathcal{B}_\infty}(Z)\|_p \leq \|E^{\mathcal{B}_n}(X) - E^{\mathcal{B}_\infty}(X)\|_p + 2\|Y\|_p.$$

Consequently, the special case above yields that

$$\|E^{\mathcal{B}_n}(X) - E^{\mathcal{B}_\infty}(X)\|_p \to 0 \quad \text{as} \quad n \to \infty.$$

Otherwise,

$$\|Y\|_p^p = \int_\Omega |Y|^p dP = \int_{Z \geq a} |Z - a|^p dP \leq \int_{Z \geq a} |Z|^p dP \to 0 \quad \text{as} \quad n \to +\infty.$$

This completes the proof. $\square$

3.6 Exercises for Chapter 3

Exercise 1.
Prove the Corollary 3.3.

Exercise 2.
(1) $E^{\mathcal{B}}(\lim_n \uparrow f_n) = \lim_n \uparrow E^{\mathcal{B}}(f_n)$.

(2) $\forall (f_n)_n$ a sequence in $\bar{L}_+$, we have $\sum_n E^{\mathcal{B}}(f_n) = E^{\mathcal{B}}(\sum_n f_n)$.

(3) $\forall (f_n)_n$ a sequence in $\bar{L}_+$, we have $E^{\mathcal{B}}(\underline{\lim}_n f_n) \leq \underline{\lim}_n E^{\mathcal{B}}(f_n)$.

Exercise 3.
Let $\mathcal{B}$ be a sub$-\sigma$-field in $\mathcal{A}$.

(1) $\forall f \in L^1$, there exists a unique element, denoted $E^{\mathcal{B}}(f) \in L^1(\mathcal{B})$, such that,

$$\int_{\Omega} E^{\mathcal{B}}(f) g dP = \int_{\Omega} f g dP, \ \forall \, g \in L^{\infty}(\mathcal{B}).$$

(2) $E^{\mathcal{B}} : L^1 \rightarrow L^1(\mathcal{B})$ is an idempotent, contractive linear operator, and satisfying $E^{\mathcal{B}}(1) = 1$.

(3) $\forall f \in L^1$, $\forall h \in L(\mathcal{B})$, such that $hf \in L^1$, we have $E^{\mathcal{B}}(hf) = hE^{\mathcal{B}}(f)$.

Exercise 4. Hölder's inequality.

Let $1 < p, q < +\infty$ be two conjugated real numbers ($\dfrac{1}{p} + \dfrac{1}{q} = 1$). Prove that $\forall f \in L^p$, $g \in L^p$,

$$|E^{\mathcal{B}}(fg)| \leq (E^{\mathcal{B}}|f|^p)^{\frac{1}{p}}(E^{\mathcal{B}}|g|^q)^{\frac{1}{q}}.$$

Exercise 5.

Let $1 \leq p, q \leq +\infty$ be such that, $\frac{1}{p} + \frac{1}{q} = 1$, and $\mathcal{B}$ be a sub-σ-field in $\mathcal{A}$. Show that

(1) $\forall f \in L^p$, there exists a unique element $E^{\mathcal{B}}(f) \in L^p(\mathcal{B})$, such that, $\forall g \in L^q$, we have

$$\int_{\Omega} E^{\mathcal{B}}(f) g dP = \int_{\Omega} f g dP.$$

(2) The map $E^p : L^q \rightarrow L^p(\mathcal{B})$ which associates to any element f its image $E^{\mathcal{B}}(f)$ is an idempotent, contractive, positive linear operator.

(3) $E^{\mathcal{B}}(1) = 1$.

(4) $\forall \, h \in L^p(\mathcal{B})$, $E^{\mathcal{B}}(hf) = hE^{\mathcal{B}}(f)$.

Exercise 6.

(1) Let $(X_n)_n$ be an integrable martingale. Show that

$$\sup_n E(X_n^+) < \infty \Longleftrightarrow \sup_n E(|X_n|) < \infty.$$

(2) Let $(X_n)_n$ be an integrable sub-martingale, such that, $\sup_n E(X_n^+) < \infty$. Prove that there exists $(X_n^1)_n$, a positive integrable martingale, (X_n^2), a positive integrable upper-martingale, satisfying $X_n = X_n^1 - X_n^2$.

(3) Show finally that for any decomposition (X_n^1, X_n^2), we have

$$X_n^1 \geq M_n, \ \ X_n^2 \geq Y_n, \ \ M_{\infty} = X_{\infty}^+ \ \text{ and } \ Y_{\infty} = X_{\infty}^-.$$

M_n, Y_n, M_{∞} and Y_{∞} are defined as in Section 3.3.

Exercise 7.

Let (X_n) be a sequence of real random variables with values in a measurable space $(E, \mathcal{F})$. For $F \in \mathcal{F}$, let ν_F be the reaching time of F by the sequence $(X_n)_n$. Denote

$$\nu_F(w) = \begin{cases} \inf\{n, \ X_n(w) \in F\} & \text{for} \quad w \in \bigcup_n \{X_n \in F\}, \\ \infty & \text{else.} \end{cases}$$

Show that ν_F is a stooping time.

Exercise 8.

Let $\nu : \Omega \to \bar{\mathbb{N}}$ be a stopping time. Show that ν is $\mathcal{B}_\infty$-measurable.

Exercise 9.

Show that for a sequence $(\nu_k)_k$ of stopping times, $\sup_k \nu_k$ and $\inf_k \nu_k$ are stopping times.

Exercise 10.

Show that Proposition 3.9 holds for $a, b \in \mathbb{Q}$ instead of $a, b \in \mathbb{R}$.

Chapter 4

Hausdorff Measure and Dimension

This chapter constitutes the basic, and the principal part of the book. In which we focus on the notion of Hausdorff measure, which is the essential measure in the heart of fractal analysis, and geometry. We develop in details the original construction of such a measure, its different variants, and the concept of the Hausdorff dimension associated to it. More information, examples, and related topics may be found in [Billingsley (1965); David and Semmes (1997); Edgar (1998, 2008); Falconer (1985, 1990); Gervais (2009); Lambert (2012); Robert (2020); Rogers (1970); Yeh (2014)].

4.1 Hausdorff measure

All the concepts that will be exposed in this chapter are considered in the Euclidean space $\mathbb{R}^d$, where $d = 1, 2, \ldots$ is an integer. Let $E \subset \mathbb{R}^d$, and $\epsilon > 0$. We call an ϵ-covering of E any Countable collection $(U_j)_j$ composed of subsets $U_j \subset \mathbb{R}^d$, satisfying

$$E \subset \bigcup U_j \quad \text{and} \quad |U_j| \leq \epsilon, \ \forall j,$$

where for a subset $U \subset \mathbb{R}^d$, $|U|$ is the diameter of U defined by

$$|U| = \sup\{\|x - y\|, \ x, y \in \mathbb{R}^d\},$$

and where for a vector $u = (u_1, u_2, \ldots, u_d) \in \mathbb{R}^d$,

$$\|u\| = \sqrt{u_1^2 + u_2^2 + \cdots + u_d^2},$$

is the usual Euclidean norm of u. The diameter of U is sometimes denoted $diam(U)$. Let next $\alpha > 0$, and denote

$$\mathcal{H}_\epsilon^\alpha(E) = \inf \sum_j |U_j|^\alpha,$$

where the lower bound is taken on the set of all ϵ-coverings of E. It is straightforward that $\mathcal{H}^\alpha_\epsilon(E)$ is monotone (non-increasing) as a function of ϵ, it has a limit in $\overline{\mathbb{R}}$. Denote

$$\mathcal{H}^\alpha(E) = \lim_{\epsilon \to 0} \mathcal{H}^\alpha_\epsilon(E), \quad \text{and} \quad \mathcal{H}^\alpha(\emptyset) = 0.$$

In the sequel, we will use the so-called Hausdorff distance of sets. Let E, F be two subsets of $\mathbb{R}^d$. For $\epsilon > 0$, denote

$$E_\epsilon = \{x \in \mathbb{R}^d; \; \|x - y\| \le \epsilon \; for \; some \; y \in E\},$$

and similarly,

$$F_\epsilon = \{x \in \mathbb{R}^d; \; \|x - y\| \le \epsilon \; for \; some \; y \in F\}.$$

The Hausdorff distance $d(E, F)$ is evaluated as

$$d(E, F) = \inf\{\epsilon > 0; \; E \subseteq F_\epsilon \; and \; F \subseteq E_\epsilon\}.$$

We will see later that this distance makes the set $\mathcal{C}(\mathbb{R}^d)$ of all compact sets in $\mathbb{R}^d$ a complete metric space.

Proposition 4.1. *For $\alpha > 0$ fixed,*

i. *$\mathcal{H}^\alpha$ is an outer metric measure on $\mathbb{R}^d$.*
ii. *$\mathcal{H}^\alpha$ is regular.*

Proof. i. We have to show that
a. $\mathcal{H}^\alpha(\varnothing) = 0$.
b. $\mathcal{H}^\alpha$ is monotone, in the sense that

$$\mathcal{H}^\alpha(E) \le \mathcal{H}^\alpha(F),$$

whenever $E \subseteq F \subseteq \mathbb{R}^n$.
c. $\mathcal{H}^\alpha$ is sub-additive, in the sense that

$$\mathcal{H}^\alpha(\bigcup_{p \ge 0} E_p) \le \sum_{p \ge 0} \mathcal{H}^\alpha(E_p).$$

for all Countable family $(E_p)_{p \ge 0}$ of subsets of $\mathbb{R}^n$.
Assertion **a.** is obvious. We shall prove **b.** Let $E \subseteq F$ be subsets of $\mathbb{R}^n$. It is straightforward that any ϵ-covering (U_j) of F is obviously an ϵ-covering of E. Consequently,

$$\mathcal{H}^\alpha_\epsilon(E) \le \sum_j |U_j|^\alpha.$$

Taking the inf on all the ϵ-covering (U_j) of F we obtain

$$\mathcal{H}^\alpha_\epsilon(E) \le \mathcal{H}^\alpha_\epsilon(F),$$

which yields, with the limit on ϵ, that

$$\mathcal{H}^\alpha(E) \leq \mathcal{H}^\alpha(F).$$

We now prove assertion **c**. Let $\epsilon > 0$, $(E_p)_p$ be a Countable family of subsets $E_p \subset \mathbb{R}^d$. Without loss of the generality, we may assume that

$$\sum_{p \geq 0} \mathcal{H}^\alpha(E_p) < \infty.$$

For $p \in \mathbb{N}$ fixed, let $(U_{p,j})_j$ be an ϵ-covering of E_p, such that,

$$\sum_j |U_{p,j}|^\alpha \leq \mathcal{H}^\alpha(E_p) + \frac{\epsilon}{2^p}.$$

It is obvious that $(U_{p,j})_{j,p}$ is an ϵ-covering of $E = \bigcup_p E_p$, and thus

$$\mathcal{H}_\epsilon^\alpha(E) \leq \sum_p \sum_j |U_{p,j}|^\alpha \leq \sum_p \bigcup_p E_p^\alpha(E_p) + \epsilon,$$

which, by letting $\epsilon \to 0$, yields assertion **c**.

We now prove that $\mathcal{H}^\alpha$ is metric. Let E, and F be subsets of $\mathbb{R}^d$, such that $d(E, F) > 0$, where d is the Hausdorff distance of sets. Let also $(U_j)_{j \in \mathbb{N}}$ be an ϵ-covering of $E \cup F$, with $0 < \epsilon < \dfrac{d(E, F)}{2}$. We immediately observe that, for all $j \in \mathbb{N}$,

$$U_j \cap E = \emptyset \quad \text{or} \quad U_j \cap F = \emptyset.$$

Therefore, for $\alpha \geq 0$, we get

$$\sum_j |U_j|^\alpha = \sum_{j;\, U_j \cap E = \emptyset} |U_j|^\alpha + \sum_{j;\, U_j \cap F = \emptyset} |U_j|^\alpha.$$

Next, as

$$\sum_{j;\, U_j \cap E = \emptyset} |U_j|^\alpha \geq \mathcal{H}_\epsilon^\alpha(E),$$

and similarly,

$$\sum_{j;\, U_j \cap F = \emptyset} |U_j|^\alpha \geq \mathcal{H}_\epsilon^\alpha(F),$$

we obtain

$$\sum_j |U_j|^\alpha \geq \mathcal{H}_\epsilon^\alpha(E) + \mathcal{H}_\epsilon^\alpha(F),$$

for all ϵ-covering $(U_j)_{j \in \mathbb{N}}$ of $E \cup F$. Consequently,

$$\mathcal{H}_\epsilon^\alpha(E \cup F) \geq \mathcal{H}_\epsilon^\alpha(E) + \mathcal{H}_\epsilon^\alpha(F), \quad \forall \epsilon,$$

which, by letting $\epsilon \downarrow 0$, yields that

$$\mathcal{H}^\alpha(E \cup F) \geq \mathcal{H}^\alpha(E) + \mathcal{H}^\alpha(F).$$

The opposite inequality is always true due to the sub-additivity property of $\mathcal{H}^\alpha$, and does not require the assumption $d(E, F) > 0$. As a result, we get

$$\mathcal{H}^\alpha(E \cup F) = \mathcal{H}^\alpha(E) + \mathcal{H}^\alpha(F) \quad \text{whenever} \quad d(E, F) > 0.$$

It remains to prove that $\mathcal{H}^\alpha$ is regular. So, let $E \subset \mathbb{R}^d$. We shall prove that there exists a $\mathcal{H}^\alpha$-measurable set $A \subset \mathbb{R}^d$, such that,

$$\mathcal{H}^\alpha(E) = \mathcal{H}^\alpha(A).$$

Observe that whenever $\mathcal{H}^\alpha(E) = \infty$, the problem is solved by choosing $A = \mathbb{R}^d$. So, the interesting case is when $\mathcal{H}^\alpha(E) < \infty$. Assume that this situation occurs, fix $n \in \mathbb{N}$, and let $\epsilon_n = \dfrac{2}{n}$. There exists an ϵ_n-covering $(V_{n,j})_j$ of E composed of open sets, and satisfying

$$\sum_j |V_{n,j}|^\alpha \leq H^\alpha_{\frac{1}{n}}(E) + \frac{1}{n}.$$

Consider next

$$A = \bigcap_n \bigcup_j V_{n,j}.$$

It is straightforward that A is a Borel set. It is therefore $\mathcal{H}^\alpha$-measurable. Moreover, $E \subset A$. Hence,

$$\mathcal{H}^\alpha(E) \leq \mathcal{H}^\alpha(A).$$

Besides,

$$A \subset \bigcup_j V_{n,j}, \quad \forall n.$$

Hence, $(V_{n,j})_j$ is an ϵ_n-covering of A. Consequently,

$$H^\alpha_{\epsilon_n}(A) \leq \sum_j |V_{n,j}|^\alpha \leq H^\alpha_{\frac{1}{n}}(E) + \frac{1}{n}, \quad \forall n.$$

As a result,

$$H^\alpha_{\epsilon_n}(A) \leq H^\alpha_{\frac{1}{n}}(E) + \frac{1}{n}, \quad \forall n.$$

Letting $n \longrightarrow \infty$, this leads to

$$\mathcal{H}^\alpha(A) \leq \mathcal{H}^\alpha(E).$$

So, finally, we get

$$\mathcal{H}^\alpha(A) = \mathcal{H}^\alpha(E).$$

$\square$

Definition 4.1. The restriction of $\mathcal{H}^\alpha$ on the σ-algebra of $\mathcal{H}^\alpha$-measurable sets is called the Hausdorff measure with dimension α, or also the α-Hausdorff measure.

Proposition 4.2. *The following assertions are true.*

(1) *For $E \subset \mathbb{R}^d$, and $0 < \alpha < \beta$, we have*

$$\mathcal{H}^\alpha(E) < \infty \implies H^\beta(E) = 0.$$

(2) *Whenever $\alpha > d$, we get $\mathcal{H}^\alpha(\mathbb{R}^d) = 0$.*

Proof 1. For any ϵ-covering $(U_j)_j$ of E, we get

$$\mathcal{H}_\epsilon^\beta(E) \leq \sum_i |U_j|^\beta \leq \epsilon^{\beta-\alpha} \sum_i |U_j|^\alpha.$$

This leads to

$$\mathcal{H}_\epsilon^\beta(E) \leq \epsilon^{\beta-\alpha}\mathcal{H}_\epsilon^\alpha(E).$$

Letting $\epsilon \downarrow 0$, we obtain

$$H^\beta(E) = 0.$$

2. We will prove that $H^\beta(C) = 0$, for any cube C of $\mathbb{R}^d$, with unit side. Indeed, write

$$C = \bigcup_{j=1}^{k^d} C_{k,j},$$

where, $C_{k,j}$ are cubes with side length $\dfrac{1}{k}$, $k \in \mathbb{N}$. Next, denote $\epsilon_k = \dfrac{\sqrt{d}}{k}$, and observe that

$$H_{\epsilon_k}^\alpha(C) \geq \sum_j |C_{k,j}|^\alpha = k^d \left(\frac{\sqrt{d}}{k}\right)^\alpha, \ \forall k.$$

Hence, letting $k \longrightarrow \infty$, we get

$$\mathcal{H}^\alpha(C) = 0.$$

$\square$

Corollary 4.1. *For all $E \subset \mathbb{R}^d$, there exists a unique critical value $\alpha_E \geq 0$, satisfying*

- $\mathcal{H}^\alpha(E) = 0, \ \forall \alpha > \alpha_E.$
- $\mathcal{H}^\alpha(E) = \infty, \ \forall \alpha < \alpha_E.$

Proof. Denote

$$\alpha_E = \inf\{\alpha > 0; \ \mathcal{H}^\alpha(E) = 0\}.$$

As the last set is not empty (it contains $d+1$), and is contained in $(0, \infty)$, so, its lower bound exists in $[0, \infty]$. Denote it by α_E. We get immediately assertions 1, and 2. □

Definition 4.2. The critical value α_E is called the Hausdorff dimension of E, and denote $dim_H E$ or simply $dimE$. Figure 4.1 illustrates the concept of the Hausdorff dimension graphically.

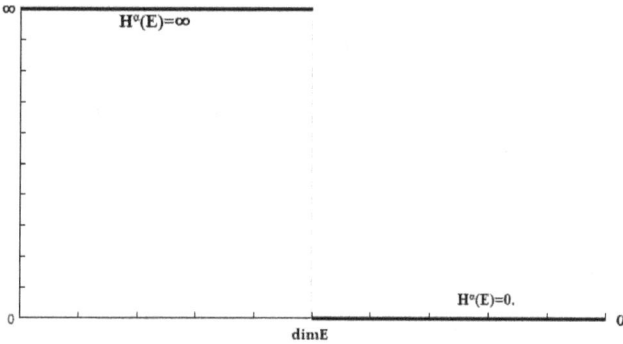

Fig. 4.1: The Hausdorff dimension of a set E.

Proposition 4.3. *The following properties hold.*

(1) $0 \le dim \, E \le d$, $\forall E \subset \mathbb{R}^d$.

(2) $dim \, E \le dim \, F$, $\forall E \subset F \subset \mathbb{R}^d$.

(3) $dim\{x\} = 0$, $\forall x \in \mathbb{R}$.

(4) $dim \bigcup_n E_n = \sup_n dim \, E_n$, $\forall (E_n)_n$ *a sequence of subsets of* $\mathbb{R}^d$.

(5) $dim \, E = 0$ *whenever* E *is a finite or Countable set in* $\mathbb{R}^d$.

(6) $\forall E \subset \mathbb{R}^d$, $0 < \mathcal{H}^\alpha(E) < \infty \implies dim \, E = \alpha$.

(7) $\forall E \subset \mathbb{R}^d$, *such that* $\lambda(E) > 0$, *we have* $dim \, E = d$, *where* λ *is the Lebesgue measure on* $\mathbb{R}^d$.

(8) $\forall E \subset \mathbb{R}^d$, $0 \le H^{dimE}(E) \le \infty$.

Proof. 1. $dim E \geq 0$ follows from its definition as a lower bound of a set of positive real numbers. On the other hand, as $E \subset \mathbb{R}^d$, and $\mathcal{H}^\alpha(\mathbb{R}^d) = 0$ for all $\alpha > d$ (See Proposition 4.2), we get $\mathcal{H}^\alpha(E) = 0$ for all $\alpha > d$. Hence, $dim E \leq d$.

2. Assume that $dim F < dim E$, and let $\eta > 0$ small enough be such that,

$$\alpha = dim F + \eta < dim E.$$

It follows that $\mathcal{H}^\alpha(E) = \infty$. Consequently, as $E \subset F$, we get $\mathcal{H}^\alpha(F) = \infty$, which contradicts the definition of $dim F$. Hence, the assertion follows.

3. $\forall x \in \mathbb{R}^d$, and $\forall \epsilon > 0$, the set $(B(x, \epsilon))$ composed of the single Ball is an ϵ-covering of $\{x\}$. Henceforth,

$$\mathcal{H}^\alpha_\epsilon(\{x\}) \leq \epsilon^\alpha, \ \forall \alpha > 0.$$

Thus,

$$\mathcal{H}^\alpha(\{x\}) = 0, \ \forall \alpha > 0.$$

Consequently, $dim\{x\} = 0$.

4. Recall that from the sub-additivity property of the Hausdorff measure, we may write, for all $\alpha > 0$,

$$\mathcal{H}^\alpha(\bigcup_n E_n) \leq \sum_n \mathcal{H}^\alpha(E_n).$$

Consequently,

$$\mathcal{H}^\alpha(\bigcup_n E_n) = 0, \ \forall \alpha > \sup_n dim E_n.$$

Which leads immediately to

$$dim(\bigcup_n E_n) \leq \sup_n dim E_n.$$

On the other hand, whenever $\alpha < \sup_n dim E_n$, there exists at least one set E_{n_0}, for some n_0, such that, $\alpha < dim E_{n_0}$. For this particular set, we have immediately $\mathcal{H}^\alpha(E_{n_0}) = \infty$. So next, as $E_{n_0} \subset \bigcup_n E_n$, we get

$$\mathcal{H}^\alpha(\bigcup_n E_n) = \infty.$$

This yields that

$$\alpha < dim(\bigcup_n E_n), \ \forall \alpha < \sup_n dim E_n.$$

Consequently,

$$\sup_n dim E_n \leq dim(\bigcup_n E_n).$$

5. This follows from assertions 3, and 4.

6. Assume that this does not occur. So, $\alpha < dimE$ or $\alpha > dimE$. In the first case, we get $\mathcal{H}^\alpha(E) = \infty$, and in the second, we get $\mathcal{H}^\alpha(E) = 0$. As both consequences are not true, we get necessarily $\alpha = dimE$.

7. For all $\epsilon > 0$, and all ϵ-covering $(U_j)_j$ of E, we get

$$\lambda(E) \leq \sum_j \lambda(U_j) \leq \sum_j |U_j|^d.$$

Consequently,

$$\lambda(E) \leq H_\epsilon^d(E), \quad \forall \epsilon > 0.$$

As a result,

$$\lambda(E) \leq H^d(E).$$

Now, as $\lambda(E) > 0$ we get $H^d(E) > 0$. Which means that $dimE \geq d$. Finally, observing assertion 1, we get $dimE = d$.

8. To prove the assertion, we shall construct sets E for which 0, and ∞ may be reached. It suffices to take $E = \mathbb{R}^d$, and $E = \mathbb{Q}^d$. □

4.2   Hausdorff dimension of Cantor-type sets

The main problem in the theory of Hausdorff measure, and dimension is the computation of the such dimension using its original definition. In the present section, we propose to evaluate the Hausdorff dimension of some particular sets known as the Cantor's type. For the sake of simplicity, we will focus on the triadic well known Cantor's set. The readers may adopt the method developed for general Cantor's sets.

The construction of the triadic Cantor set starts from the unit interval $I_0 = [0, 1]$. Next, in a first step we split I_0 into three intervals,

$$I_{00} = [0, \frac{1}{3}], \quad I_{01} = [\frac{1}{3}, \frac{2}{3}], \quad \text{and } I_{02} = [\frac{2}{3}, 1].$$

We keep next the first, the third ones, and delete the middle one. We get so, two intervals

$$I_{00} = [0, \frac{1}{3}], \quad \text{and } I_{02} = [\frac{2}{3}, 1].$$

Next, at the second step we split each one of the last intervals into three sub-intervals always with the same length. We get for I_{00},

$$I_{000} = [0, \frac{1}{9}], \quad I_{001} = [\frac{1}{9}, \frac{2}{9}], \quad \text{and } I_{002} = [\frac{2}{9}, \frac{3}{9}],$$

and similarly for I_{02},

$$I_{020} = [\frac{6}{9}, \frac{7}{9}], \quad I_{021} = [\frac{7}{9}, \frac{8}{9}], \quad \text{and} \quad I_{022} = [\frac{8}{9}, 1].$$

Keep next the first, the third ones, and delete the middle for both subdivisions of I_{00}, and I_{02}. We get so, 4 intervals,

$$I_{000} = [0, \frac{1}{9}], \quad I_{002} = [\frac{2}{9}, \frac{3}{9}], \quad I_{020} = [\frac{6}{9}, \frac{7}{9}], \quad \text{and} \quad I_{022} = [\frac{8}{9}, 1].$$

We continue the process similarly. So, given an integer $n \in \mathbb{N}$, we get at the step n a number 2^n of intervals with the same length $\frac{1}{3^n}$ of the form

$$I_{n,k} = [\frac{a_k}{3^n}, \frac{a_k + 1}{3^n}],$$

for some integer a_k. Figure 4.2 below illustrates the process.

Fig. 4.2: The triadic Cantor set.

Next, denote for $n \in \mathbb{N}$,

$$E_n = \bigcup_{k=1}^{2^n} I_{n,k}.$$

Definition 4.3. The triadic Cantor set is defined by $C = \bigcap_{n \geq 0} E_n$.

The following are special characteristics of the triadic Cantor set C, and give a first example of exact calculus in fractal geometry.

Lemma 4.1. *For all $k \in \mathbb{N}$, the set E_k is the union of 2^k intervals*

$$I_{a_k} = [\frac{a_k}{3^k}, \frac{a_k + 1}{3^k}],$$

where a_k is of the form

$$a_k = \sum_{i=0}^{k} x_i 3^{k-i}, \quad x_i \in \{0, 2\}, \quad \forall i.$$

Proof. We proceed by recurrence on the integer k. Indeed, for $k = 0$, $E_0 = [0, 1]$, let $a_0 = 0$, and $I_{a_0} = E_0$. Assume next that E_k is a union of 2^k intervals

$$I_{a_k} = [\frac{a_k}{3^k}, \frac{a_k + 1}{3^k}].$$

The next step in the construction of the triadic Cantor set consists of subdividing each interval of E_k into 3 sub- intervals with the same length, and omitting the middle one. This means that I_{a_k} is transformed into a union $I_{a_k}^1 \cup I_{a_k}^2$, where

$$I_{a_k}^1 = [\frac{a_k}{3^{k+1}}, \frac{a_k + 1}{3^{k+1}}],$$

and

$$I_{a_k}^2 = [\frac{a_k + 2.3^k}{3^{k+1}}, \frac{a_k + 2.3^k + 1}{3^{k+1}}].$$

We thus get E_{k+1} as a union of 2^{k+1} intervals of the form above with $a_{k+1} = a_k$, or $a_{k+1} = a_k + 2.3^k$, which guaranties that for a_{k+1} also the x_i's are 0 or 2.

Corollary 4.2.

$$C = \left\{ \sum_{i=1}^{\infty} \frac{a_i}{3^i}; \ a_i \in \{0, 2\}, \ \forall i \right\}.$$

Indeed, this follows from the last Lemma. Let $x \in C$. So, $x \in E_k$ for all k. Consequently, there exists a sequence

$$a_k = \sum_{i=0}^{k} x_i 3^{k-i}, \quad x_i \in \{0, 1, 2\}, \quad \forall i.$$

such that, $x \in I_{a_k}$. Which yields otherwise that

$$\frac{a_k}{3^k} = \sum_{i=0}^{k} \frac{x_i}{3^i} \leq x \leq \frac{a_k+1}{3^k} = \sum_{i=0}^{k} \frac{x_i}{3^i} + \frac{1}{3^k}.$$

Letting $k \to \infty$ we get

$$x = \sum_{i=0}^{+\infty} \frac{x_i}{3^i}.$$

The following proposition resumes some topological properties of the triadic Cantor set C.

Proposition 4.4. *The triadic Cantor set C is non-empty, compact, perfect, with empty interior, non Countable, and with Lebesgue measure zero.*

Proof. i. C is non empty as it contains 0, and 1.

ii. C is closed as $C = \bigcap\limits_{k \geq 0} E_k$, which is an intersection of closed intervals. It is obviously bounded as $C \subset [0,1]$. So; it is compact.

iii. The interior of C is empty. We proceed by the converse. So, whenever C contains an interval of length $\epsilon > 0$, such interval will be surely contained in E_k, $\forall\, k$, and consequently contained in one of the intervals I_{a_k}, $\forall\, k$. As a result,

$$0 < \epsilon < \frac{1}{3^k}, \ \forall k,$$

which is contradictory.

iv. C is perfect. Here also we use the converse reasoning. Assume that C is not perfect. Hence, there exists real numbers $a < b$ such that $[a,b] \subset C$. So,

$$[a,b] \subset E_k, \ \forall\, k \geq 0.$$

Thus, there exists a_k such that $[a,b] \subset I_{a_k}$. This yields that

$$b - a \leq \left(\frac{1}{3}\right)^k \forall\, k,$$

which is impossible.

v. C is non Countable. Indeed, if it is Countable, we may write it in a Countable way $C = \{x_1, x_2, \ldots, x_n, \ldots\}$. We will construct another element of C which is different from all these x_k's. To do it, we consider the triadic representation of the elements $x_1, \ldots, x_n, \ldots$. Let next $x = a_1, a_2, \ldots, a_n, \ldots$ be the triadic real number constructed as follows. The element $a_1 \in \{0, 2\}$ is chosen to be different from the first digit after the decimal point of x_1. It is always possible as the digits $x_{i,1}$ of x_1 are 0 or 2. If $x_{1,1} = 0$, we take $a_1 = 2$, and for $x_{1,1} = 2$, we take $a_1 = 0$. Next, we chose the digit $a_2 \in \{0, 2\}$ to be different from the second digit after the decimal point of x_2, and so on. We obtain, consequently, an element $x \in C$ different from all the elements x_k's, which is contradictory

vi. Denote $\mathfrak{m}$ the Lebesgue measure on $\mathbb{R}$. We have

$$\mathfrak{m}(C) \leq \mathfrak{m}(E_k) \leq \underset{\mathfrak{m}}{\sum} (I_{a_k}) = \sum_{i=0}^{2^k-1} (\frac{1}{3})^k = (\frac{2}{3})^k \longrightarrow 0, \quad \text{as} \quad k \to \infty.$$

$\square$

Theorem 4.1.

$$dim\, C = \frac{\log 2}{\log 3}.$$

Proof. Denote $\alpha = \dfrac{\log 2}{\log 3}$. In one hand, we have $C \subset E_n, \; \forall n$. Consequently, $(I_{n,k})_{1 \leq k \leq 2^n}$ is an ϵ_n-covering of C, where $\epsilon_n = \dfrac{1}{3^n}$. Hence,

$$\mathcal{H}^\alpha_{\epsilon_n}(C) \leq \sum_{k=1}^{2^n} |I_{n,k}|^\alpha = 2^n \left(\frac{1}{3^n} \right)^\alpha = 1.$$

As a result,

$$\mathcal{H}^\alpha(C) \leq 1.$$

As a consequence,

$$\alpha \geq dim\, C.$$

We now prove the opposite inequality. Let $\gamma, \epsilon \in \mathbb{R}$ be such that

$$\gamma > 1, \; 0 < \epsilon < \frac{1}{3} \text{ and } \gamma\epsilon < 1.$$

Let also $(U_j)_j$ be an ϵ-covering of C, and consider next a sequence $(L_j)_j$ of open intervals, such that,

$$U_j \subset L_j \text{ and } |L_j| = \gamma|U_j|, \; \forall j.$$

For any integer $p \in \mathbb{N}$, we need to estimate the number of intervals $I_{p,k}$ which intercept L_j. Denote N_j^p such a number. Remark next that, for all j, there exists a unique integer $n = n_j$ such that

$$\frac{1}{3^{n+1}} \leq |L_j| < \frac{1}{3^n}.$$

Thus, the interval L_j intercepts E_n, and more precisely, it intercepts only one interval $I_{n,k}$ from the 2^n intervals composing E_n. Hence, $N_j^n = 1$. Next, whenever $p > n$, L_j intercepts at most 2^{p-n} interval from the 2^p intervals of the step p. Consequently,

$$N_j^p \leq 2^{p-n}, \quad \forall p \geq n_j.$$

Observing that

$$2^{p-n} = 2^{p+1} \left(\frac{1}{3^{n+1}} \right)^\alpha \leq 2^{p+1} |L_j|^\alpha,$$

we get

$$N_j^p \leq 2^{p+1} |L_j|^\alpha, \quad \forall p \geq n_j.$$

Consequently,

$$2^p \leq \sum_j N_j^p \leq 2^{p+1} \sum_j |L_j|^\alpha.$$

Therefore,

$$\sum_j |L_j|^\alpha \geq \frac{1}{2}.$$

As a result,

$$\gamma^\alpha \sum_j |U_j|^\alpha \geq \frac{1}{2}.$$

This leads to

$$\mathcal{H}_\epsilon^\alpha(C) \geq \frac{1}{2\gamma^\alpha}.$$

Consequently,

$$\mathcal{H}^\alpha(C) \geq \frac{1}{2\gamma^\alpha}.$$

So that,

$$\alpha \leq \dim C.$$

$\square$

4.3   Other variants of Hausdorff dimension

In this section, some equivalent definitions of the Hausdorff dimension of sets are discussed, for which it is not necessary to consider general coverings. We show that some restrictions on special coverings to compute the Hausdorff dimension are possible. This is important as it will permit the exact, and direct computation of the Hausdorff dimension from its definition which is usually the most difficult task in the theory of Hausdorff dimension of sets.

Let $\mathcal{F}$ be a family of subsets of $\mathbb{R}^n$, satisfying

$$\forall E \subset \mathbb{R}^n, \ \forall \epsilon > 0, \ \exists \ \epsilon\text{-covering of } E \text{ by elements from } \mathcal{F}. \qquad (4.1)$$

For $E \subset \mathbb{R}^n$, and $\alpha, \epsilon > 0$, denote

$$\mathcal{H}_\epsilon^\alpha(E, \mathcal{F}) = \inf \sum_j |U_j|^\alpha,$$

where the lower bound is taken over all the ϵ-coverings of E by elements $(U_j)_j \subset \mathcal{F}$. As for the case of the Hausdorff measure, we get here a Decreasing function relatively to ϵ. So, denote

$$\mathcal{H}^\alpha(E, \mathcal{F}) = \lim_{\epsilon \to 0} \mathcal{H}_\epsilon^\alpha(E, \mathcal{F}).$$

It consists here-also of an outer, and regular metric measure $\mathcal{H}^\alpha(., \mathcal{F})$ on $\mathbb{R}^d$. It yields a dimension for sets as

$$\dim_{\mathcal{F}} E = \inf \left\{ \alpha > 0; \mathcal{H}^\alpha(E, \mathcal{F}) = 0 \right\}.$$

Definition 4.4. The collection $\mathcal{F}$ is said to permit the computation of the Hausdorff dimension of sets in $\mathbb{R}^d$ iff

$$\dim_{\mathcal{F}} E = \dim E, \ \forall E \subset \mathbb{R}^n.$$

We will expose now an example of collections $\mathcal{F}$ of subsets of $\mathbb{R}^n$ permitting the computation of the Hausdorff dimension.

Example 4.1. Consider on $[0, 1[$ a collection $\mathcal{F}$ of sub-intervals, satisfying

a. $\mathcal{F} = \bigcup_{n \geq 0} \mathcal{F}_n$.
b. $\forall n$, $\mathcal{F}_n$ is a finite partition of $[0, 1[$ on semi-open intervals closed at the left.
c. $\mathcal{F}_{n+1}$ is a refinement of $\mathcal{F}_n$, in the sense that, any element $I \in \mathcal{F}_{n+1}$ is strictly contained in one element $p(I) \in \mathcal{F}_n$, called its father or predecessor.

d. $\forall x \in [0, 1[,\ \forall \epsilon > 0$ there exists $I \in \mathcal{F}$ such that $x \in I$, and $|I| \leq \epsilon$.

e. $\forall \alpha > 0$, we have

$$\limsup_{\epsilon \to 0} \left\{ |I|^\alpha K(I),\ I \in \mathcal{F},\ |I| \leq \epsilon \right\} \leq 1,$$

where

$$K(I) = \sup\left\{ \frac{|I|}{|J|},\ J \in \mathcal{F},\ p(J) = I \right\}.$$

Hence, the collection $\mathcal{F}$ permits the computation of the Hausdorff dimension on $[0, 1[$. Otherwise,

$$\dim_{\mathcal{F}} E = \dim E,\ \forall E \subset [0, 1[.$$

Indeed, let $I \subset [0, 1[$ be such that,

$$|I| < \inf\left\{ |J|;\ J \in \mathcal{F}_0 \right\}.$$

One of the following assertions holds.

i. There exists $J \in \mathcal{F}$ such that $J \subset I \subset p(J)$.

ii. There exists $J_1, J_2 \in \mathcal{F}$ disjoint such that

$$J_1 \bigcup J_2 \subset I \subset p(J_1) \bigcup p(J_2).$$

Indeed, let $J_1 \in \mathcal{F}$ such that $J_1 \subset I$. Whenever $p(J_1) \not\supset I$, and J_1 has a minimal order k, and thus $J_1 \in \mathcal{F}_k$ consider $H \in \mathcal{F}_{k-1}$ contiguous to $p(J_1)$. There exists J_2 contiguous to $p(J_1)$, such that, $J_2 \subset I$, and $p(J_2) \not\subset I$. We immediately get

$$J_1 \bigcup J_2 \subset I \subset p(J_1) \bigcup p(J_2).$$

So, assertion **ii** holds. Next, observe that

$$\forall \epsilon > 0,\ \exists \eta > 0,\ \text{such that,}\ \forall I \in \mathcal{F},\ |I| < \eta \Rightarrow p(I) < \epsilon. \tag{4.2}$$

Denote $a_n = \inf\{|I|, I \in \mathcal{F}_n\}$. It is straightforward that $a_n \searrow 0$ as $n \to \infty$. On the other hand,

$$a_n \geq l > 0\ \forall n.$$

Denote

$$E_n = \bigcup\{I \in \mathcal{F}_n;\ |I| \geq l\}.$$

It consists of a decreasing sequence of sets. Let next $E = \bigcap E_n$. As $\mathfrak{m}(E_n) \geq l$, it holds immediately that $\mathfrak{m}(E) > 0$. Therefore, there exists $t \in [0, 1[,$

such that, $|I_n(t)| > 0$, which is contradictory. Otherwise, $\forall \alpha > 0$, there exists M_α, such that,

$$|I|^\alpha K(I) < M_\alpha, \ \forall I \in \mathcal{F}.$$

Now, for $\epsilon > 0$, and η defined in (4.2), denote $E = \bigcup_j I_j$, where I_j are intervals contained in $[0, 1[$. Denote also

$$\mathcal{L}_1 = \{I_j; \ I_j \text{ satisfies assertion i.}\},$$

and

$$\mathcal{L}_2 = \{I_j; \ I_j \text{ satisfies assertion ii.}\}.$$

Whenever $I \in \mathcal{L}_1$, there exists $J \in \mathcal{F}$, such that, $J \subset I \subset p(J)$. Similarly, whenever $I \in \mathcal{L}_2$, there exists J_1, J_2 disjoint in $\mathcal{F}$, satisfying

$$J_1 \bigcup J_2 \subset I \subset p(J_1) \bigcup p(J_2),$$

and

$$E \subset (\bigcup p(J)) \bigcup (\cup p(J_1)) \bigcup (\cup p(J_2)).$$

Now observe that $I \in \mathcal{L}_1$, and $|J|, |J_1|, |J_2| < \eta$, for $\eta > 0$ small enough. Consequently,

$$|p(J)|, |p(J_1)|, |p(J_2)| < \epsilon.$$

We thus get an ϵ-covering of E by elements of $\mathcal{F}$. Moreover, for all $\beta > 0$, we have

$$\begin{aligned}
\sum_{I \in \mathcal{L}_1} |p(J)|^{\alpha(1+\beta)} &= \sum_{I \in \mathcal{L}_1} |p(J)|^\alpha |p(J)|^{\alpha\beta} \\
&= \sum_{I \in \mathcal{L}_1} \left(|p(J)|^\beta \frac{|p(J)|}{|J|} \right)^\alpha |J|^\alpha \\
&= M_\beta^\alpha \sum_{I \in \mathcal{L}_1} |I|^\alpha.
\end{aligned}$$

Similarly, we get

$$\sum_{I \in \mathcal{L}_2} |p(J_1)|^{\alpha(1+\beta)} \leq M_\beta^\alpha \sum_{I \in \mathcal{L}_2} |I|^\alpha,$$

and

$$\sum_{I \in \mathcal{L}_2} |p(J_2)|^{\alpha(1+\beta)} \leq M_\beta^\alpha \sum_{I \in \mathcal{L}_2} |I|^\alpha.$$

As a result of these estimations, we obtain

$$\sum_{I \in \mathcal{L}_1 \cup \mathcal{L}_2} |p(J)|^{\alpha(1+\beta)} \leq 3 M_\beta^\alpha \sum_I |I|^\alpha.$$

This yields that

$$\mathcal{H}_\epsilon^{\alpha(1+\beta)}(E, \mathcal{F}) \leq 3M_\beta^\alpha \mathcal{H}_\eta^\alpha(E).$$

Letting $\epsilon \downarrow 0$, this implies that

$$\mathcal{H}_\epsilon^{\alpha(1+\beta)}(E, \mathcal{F}) = 0.$$

Therefore,

$$\dim_{\mathcal{F}} E \leq \alpha(1 + \beta), \quad \forall \beta > 0.$$

Consequently,

$$\dim_{\mathcal{F}} E \leq \alpha.$$

4.4   Upper and lower bounds of the Hausdorff dimension

In this part, we propose to present an upper bound for the Hausdorff dimension of sets. Let E be a bounded subset in $\mathbb{R}^d$. For $\epsilon > 0$, denote $N_\epsilon(E)$ the minimum number of Balls with diameter ϵ that cover E. Denote also

$$\delta(E) = \liminf_{\epsilon \searrow 0} \frac{\log N_\epsilon(E)}{-\log \epsilon}.$$

The following result shows an upper bound of the Hausdorff dimension of sets.

Proposition 4.5. *For all bounded set $E \subset \mathbb{R}^d$, we have*

$$dim E \leq \delta(E).$$

Proof. Let $\eta > 0$, and $\alpha > \delta(E)$. Consider also a collection of Balls $(B_j)_{1 \leq j \leq N_\epsilon(E)}$, such that, $|B_j| = \epsilon \leq \eta$. It holds that

$$\mathcal{H}_\eta^\alpha(E) \leq \sum_j |B_j|^\alpha = N_\epsilon(E)\epsilon^\alpha. \quad \forall \epsilon, \, 0 < \epsilon < \eta.$$

Next, as $\delta(E) < \alpha$, we get $N_\epsilon(E)\epsilon^\alpha \leq 1$, for some $\eta > 0$ small enough, and for all ϵ, $0 < \epsilon < \eta$. Consequently, $\mathcal{H}_\eta^\alpha(E) \leq 1$, which yields that $\mathcal{H}^\alpha(E) < \infty$. Hence,

$$dim E \leq \alpha, \, \forall > \delta(E).$$

Thus, $dim E \leq \delta(E)$. $\square$

Now, similarly to the previous case, we aim to give a lower bound for the Hausdorff dimension of sets. To do this we introduce some useful concepts that will be used.

Definition 4.5. A Borel measure μ on $\mathbb{R}^d$ is said to be α-Hölder ($\alpha \geq 0$) iff there exists a constant $C > 0$, such that, $\mu(B) \leq C|B|^\alpha$, for any Ball B in $\mathbb{R}^d$. We say also that μ is α-Hölderian, or is Hölderian with exponent or index α.

The following proposition shows a lower bound for the Hausdorff dimension of sets in some special cases.

Proposition 4.6. *Let E be a Borel subset in $\mathbb{R}^d$, for which, there exists a Hölderian measure μ of index α on $\mathbb{R}^d$, such that, $\mu^*(E) > 0$. Then, $\mathcal{H}^\alpha(E) > 0$, and consequently, $\dim E \geq \alpha$, where*

$$\mu^*(E) = \inf\left\{\sum_j \mu(U_j); \ U_j \text{ is bounded } \forall j, \text{ and } \bigcup_j U_j \supset E\right\}.$$

Proof. Consider for $\epsilon > 0$ an ϵ-covering of E by means of Balls $(B_j)_j$. It holds that

$$\frac{1}{C}\mu^*(E) \leq \frac{1}{C}\sum_j \mu(B_j) \leq \sum_j |B_j|^\alpha.$$

This yields that

$$0 < \frac{1}{C}\mu^*(E) \leq \mathcal{H}^\alpha(E, \mathcal{F}) \leq 2^\alpha \mathcal{H}^\alpha(E),$$

where $\mathcal{F}$ is the collection of Balls. Consequently, $\mathcal{H}^\alpha(E) > 0$, which implies that $\dim E \geq \alpha$. $\qquad\square$

Proposition 4.7. *Let $E \subset \mathbb{R}^d$, $\alpha \geq 0$, and $\Phi : \mathbb{R}^d \to \mathbb{R}^d$ be such that,*

$$|\Phi(x) - \Phi(y)| \leq C|x - y|^\alpha, \ \forall x, y \in \mathbb{R}^d.$$

Then,

$$\dim\Phi(E) \leq \frac{1}{\alpha}\dim E.$$

The function Φ above is said to be α-Hölder or α-Hölderian or Hölderian with exponent or index α.

Proof. Denote $\gamma = \dim E$, and for $\eta > 0$, denote

$$s = \frac{\gamma}{\alpha}(1 + \eta).$$

Consider next, for $\epsilon > 0$, an ϵ-covering $(I_j)_j$ of E, and denote $\delta = C\epsilon^\alpha$. It holds immediately that

$$|\Phi(I_j)| \leq C|I_j|^\alpha.$$

This yields that $(\Phi(I_j))_j$ is a δ-covering of $\Phi(E)$. As a result

$$\mathcal{H}^s_\delta(\Phi(E)) \leq C^s \mathcal{H}^{\gamma(1+\eta)}_\epsilon(E).$$

Letting $\epsilon \downarrow 0$, we obtain

$$\mathcal{H}^s(\Phi(E)) = 0,$$

which means that $s \geq \dim\Phi(E)$. Letting now $\eta \downarrow 0$, we obtain

$$\frac{\gamma}{\alpha} \geq \dim\Phi(E).$$

$\square$

4.5 Frostman's Lemma

It is a basic result in measure theory, which is applied widely in multi-fractal analysis. It permits to construct measures of Frostman's type on multifractal sets whenever Gibbs hypothesis is no longer valid.

Theorem 4.2. *Let $A \subset \mathbb{R}^d$ be a compact set, $\alpha \geq 0$, such that, $\mathcal{H}^\alpha(A) > 0$. Then, there exists a Borel probability measure μ on $\mathbb{R}^d$, supported on A, and satisfying*

$$\mu(B) \leq M|B|^\alpha,$$

for all Ball B in $\mathbb{R}^d$. $\mathcal{H}^\alpha$ being the α-Hausdorff measure.

Proof. We will split the proof into steps.
Claim 1. There exists a constant $\gamma > 0$ such that

$$\sum_j |U_j|^\alpha \geq \gamma,$$

for all covering $(U_j)_j$ of A.
Claim 2. There exists a sequence $(\mu_n)_n$ of Borel probability measures on $\mathbb{R}^d$, such that,

$$\mu_n(C) \leq M|C|^\alpha,$$

for all dyadic cube C of order $\leq n$, and

$$\mu_n(E_n) = 1,$$

where E_n is the union of all dyadic cubes of order n intersecting A.

Claim 3. The sequence $(\mu_n)_n$ is relatively compact, so that, it has a subsequence $(\mu_{n_k})_k$ converging weakly to a measure μ.

Proof of claim 1. Observe that

$$\mathcal{H}^\alpha(A) = \lim_{\epsilon \downarrow 0} \mathcal{H}^\alpha_\epsilon(A).$$

As a result, there exists $\epsilon > 0$, such that, $\mathcal{H}^\alpha_\epsilon(A) > 0$. Let $(U_j)_j$ be a covering of A. We have obviously

$$\sup_j |U_j| \leq \epsilon \text{ or } \sup_j |U_j| > \epsilon.$$

Consequently,

$$\sum_j |U_j|^\alpha \geq \min\{\epsilon^\alpha, \mathcal{H}^\alpha_\epsilon(A)\} = \gamma.$$

Proof of claim 2. Let $n \in \mathbb{N}$, be fixed, and consider the dyadic cubes of order n intersecting A. Consider next the measure μ^n_n defined by

$$\mu^n_n(C) = (\frac{1}{2^n})^\alpha \quad \text{if} \quad C \cap A \neq \emptyset,$$

and 0 iff $C \cap A = \emptyset$. Next, at the order $n-1$, the weights will be evaluated according to μ^n_n. For a dyadic cube C of order $(n-1)$ intersecting A, we put

$$\mu^{n-1}_n(C) = \mu^n_n(C) \text{ whenever } \mu^n_n(C) \leq (\frac{1}{2^{n-1}})^\alpha.$$

When the same cube is such that $\mu^n_n(C) > (\frac{1}{2^{n-1}})^\alpha$, we put

$$\mu^{n-1}_n(C) = \lambda \mu^n_n(C),$$

where $0 < \lambda < 1$ is such that,

$$\mu^{n-1}_n(C) = (\frac{1}{2^{n-1}})^\alpha.$$

Of course, when $C \cap A = \emptyset$, we put $\mu^{n-1}_n(C) = 0$. We thus get

$$\begin{cases} \mu^{n-1}_n(C_n) \leq (\frac{1}{2^n})^\alpha \text{ if } C \text{ is a dyadic cube of order } n. \\ \mu^{n-1}_n(C) \leq (\frac{1}{2^{n-1}})^\alpha \text{ if } C \text{ is a dyadic cube of order } n-1. \end{cases}$$

We follow the process until we get μ_n^1. This latter is a finite measure on $\mathbb{R}^d$, and moreover, it satisfies

$$\mu_n^1(C) \leq (\frac{1}{2^k})^\alpha, \, \forall C \text{ a dyadic cube of order } k \leq n,$$

and

$$\mu_n^1(E_n) = ||\mu_n^1|| < \infty,$$

where E_n is the union of all dyadic cubes of order n intersecting A. We put next

$$\mu_n = \frac{1}{||\mu_n^1||}\mu_n^1.$$

Observe next that $\forall t \in A$, there exists a dyadic cube C of order $k \leq n$, such that, $t \in C$, and

$$\mu_n^1(C) = (\frac{1}{2^k})^\alpha,$$

which may be written as

$$\mu_n^1(C) = (\frac{1}{\sqrt{d}})^\alpha \left(\frac{\sqrt{d}}{2^k} \right)^\alpha.$$

Denoting $\beta = (\frac{1}{\sqrt{d}})^\alpha$, and observing that $|C| = \frac{\sqrt{d}}{2^k}$, we get

$$\mu_n^1(C) = \beta|C|^\alpha.$$

Next, for $t \in A$, let C_t be the largest dyadic cube of order $\leq n$ containing t, and satisfying

$$\mu_n(C_t) = \beta|C_t|^\alpha.$$

We immediately observe that whenever $t' \neq t$ in A, we get either

$$C_t = C_{t'} \text{ or } C_t \cap C_{t'} = \emptyset.$$

Hence, as A is compact, and $A \subset \cup_{t\in A}C_t$, we may cover it with a finite number p of cubes C_{t_j}, $1 \leq j \leq p$ (the cubes being dyadic, with orders $\leq n$, and disjoint). Consequently,

$$||\mu_n^1|| \geq \mu_n^1(\bigcup_j C_{t_j}) = \sum_j \mu_n^1(C_{t_j}) = \beta \sum_j |C_{t_j}|^\alpha \geq \beta\gamma.$$

As a result,

$$\mu_n(C) \leq \frac{1}{\beta\gamma}\beta|C|^\alpha = \frac{1}{\gamma}|C|^\alpha$$

for all dyadic cube C of order $\leq n$. Furthermore, $\mu_n(E_n) = 1$. Denote next, $\Gamma = \{\mu_n; \ n \in \mathbb{N}\}$. It is straightforward that

$$E_n \subset E_{n-1} \subset \cdots \subset E_1.$$

Consequently,

$$\mu_n(\overline{E_1}) = 1, \ \forall n.$$

Hence, Γ is tight. Consequently, it is relatively compact. Hence, there exists a sub-sequence $(\mu_{n_k})_k$ of $(\mu_n)_n$ converging in the weak sense to a Borel measure μ on $\mathbb{R}^d$. We will prove now that μ is α-Hölderian. Let

$$C = \prod_{j=1}^{d} [\frac{m_j}{2^n}, \frac{m_j + 1}{2^n}[, \quad m_j \in \mathbb{Z},$$

be a dyadic cube of order n, and let

$$\tilde{C} = \prod_{j=1}^{d} [\frac{m_j - 1}{2^n}, \frac{n_j + 2}{2^n}[.$$

Let next $f : \mathbb{R}^d \longrightarrow [0, 1]$ be continuous, such that $f \equiv 1$ on C, support$(f) \subset \tilde{C}$, and $0 \leq f \leq 1$. We have

$$\mu(C) \leq \int_C f d\mu = \lim_{k \to +\infty} \int_{\tilde{C}} f d\mu_{n_k} \leq \lim_{k \to +\infty} \mu_{n_k}(\tilde{C}).$$

As $\tilde{C}$ is a union of 3^d dyadic cubes of order n of type C, we get

$$\mu(C) \leq 3^d \frac{1}{\gamma} |C|^\alpha.$$

Let now $B(x, r)$ be a Ball in $\mathbb{R}^d$. If $2r < 1$, let n be such that,

$$\frac{1}{2^n} \leq 2r \leq \frac{1}{2^{n-1}}.$$

$B(x, r)$ is contained in at most 3^d dyadic cubes C_j of order n. Consequently,

$$\mu(B(x, r)) \leq 3^d \mu(C_j) \leq 3^d 3^d \frac{1}{\gamma} |C_j|^\alpha = \frac{9^d}{\gamma} (\sqrt{d})^\alpha |2r|^\alpha.$$

Denoting $M = \frac{9^d}{\gamma} (\sqrt{d})^\alpha$, this means that

$$\mu(B(x, r)) \leq M |2r|^\alpha.$$

It remains finally to show that $\mu(A) = 1$. Indeed, as $A = \cap_{n=1}^{\infty} \overline{E_n}$, we get

$$\mu(A) = \lim_{n \to +\infty} \mu(\overline{E_n}).$$

On the other hand, we have

$$\mu(\overline{E_n}) \geq \limsup_{k \to +\infty} \mu_{n_k}(\overline{E_n}) = 1.$$

Consequently, $\mu(A) = 1$.

4.6   Application

Let $\{b_j\} \subset \mathbb{N}$, $b_j \geq 2$, and denote $a_j = \prod_{1 \leq k \leq j} b_k$. For all $x \in [0,1]$, consider its expression as

$$x = \sum_{j \geq 1} \frac{x_j}{a_j} \text{ with } x_j \in \{0, 1, 2, \ldots, b_{j-1}\}.$$

Consider next the sets

$$E = \left\{ x = \sum_{j \geq 0} \frac{\epsilon_{2j+1}}{a_{2j+1}}, \ 0 \leq \epsilon_{2j+1} < b_{2j+1} \right\},$$

and

$$F = \left\{ x = \sum_{j \geq 0} \frac{\epsilon_{2j}}{a_{2j}}, \ 0 \leq \epsilon_{2j} < b_{2j} \right\}.$$

Then E, and F are compact. We now show that for a suitable choice of (b_j), we may obtain

$$\dim E = \dim F = 0.$$

For $n \in \mathbb{N}$, consider the collection of intervals in $[0, 1[$ defined by

$$\mathcal{F}_n = \left\{ I_n = \left[\sum_{j=1}^{n} \frac{x_j}{a_j}, \sum_{j=1}^{n} \frac{x_j}{a_j} + \frac{1}{a_n} \right] \right\}.$$

It is easy to see that $\{\mathcal{F}_n\}_n$ is a partition of $[0, 1[$, and that for all n, $\mathcal{F}_{n+1}$ is a refinement of $\mathcal{F}_n$. Moreover, for $0 \leq x < 1$, there exists n, such that, $x \in I_n$. Denote as previously

$$E = \left\{ \sum_{j} \frac{\epsilon_j}{a_j}, \ 0 \leq \epsilon_j < b_{j+2p} = 0 \right\},$$

and

$$F = \left\{ \sum_{j} \frac{\epsilon_j}{a_j}, \ 0 \leq \epsilon_j < b_{j+2p+1} = 0 \right\}.$$

We obtain here compact subsets E, and F in $[0, 1[$. Let next

$$\mathbb{A} = \{ \epsilon = (\epsilon_n), \epsilon_n \in \{0, 1, 2, \ldots, b_{n-1}\} \},$$

and d be the distance defined on $\mathbb{A}$ as follows,

⋆ $d(\epsilon, \epsilon') = 0$ whenever $\epsilon = \epsilon'$.
⋆ $d(\epsilon, \epsilon') = 1$ whenever $\epsilon_1 \neq \epsilon'_1$ (and thus $\epsilon \neq \epsilon'$).
⋆ $d(\epsilon, \epsilon') = \dfrac{1}{a_n}$ whenever $\epsilon_1 = \epsilon'_1 \ldots, \epsilon_n = \epsilon'_n$, and $\epsilon_{n+1} \neq \epsilon'_{n+1}$.

The map d constructed above is an ultra-metric distance, as it satisfies particularly

$$d(\epsilon, \epsilon') \leq \sup\{d(\epsilon, \epsilon''), d(\epsilon'', \epsilon')\}.$$

We may show that

i. $(\mathbb{A}, d)$ is compact.
ii. The sets

$$\widetilde{E} = \{\epsilon = (\epsilon_j)/\epsilon_{2p} = 0\},$$

and

$$\widetilde{F} = \{\epsilon = (\epsilon_j)/\epsilon_{2p+1} = 0\}$$

are closed in $\mathbb{A}$, and thus compact.

Consider the function

$$\varphi : \mathbb{A} \to [0, 1]$$
$$\epsilon \mapsto \sum_{j \geq 1} \frac{\epsilon_j}{a_j}.$$

We immediately notice that

$$\varphi(\widetilde{E}) = E,$$

and

$$\varphi(\widetilde{F}) = F.$$

Consequently, it suffices to show that φ is continuous, particularly,

$$|\varphi(\epsilon) - \varphi(\epsilon')| \leq d(\epsilon, \epsilon').$$

For $\epsilon = \epsilon'$, the inequality is satisfied. For $\epsilon_1 \neq \epsilon_1'$, the inequality is also satisfied. Whenever $\epsilon_1 = \epsilon_1', \ldots, \epsilon_n = \epsilon_n'$, and $\epsilon_{n+1} \neq \epsilon_{n+1}'$, we have

$$|\varphi(\epsilon) - \varphi(\epsilon')| \leq \frac{1}{a_n} = d(\epsilon, \epsilon').$$

We now recall that

$$\dim E \leq \delta(E) = \liminf_{\epsilon \to 0} \frac{\log N_\epsilon(E)}{-\log \epsilon}.$$

On the other hand,

$$N_{\frac{1}{a_{2k}}}(E) \leq \prod_{j=1}^{k} b_{2j-1}.$$

Consequently,

$$\dim E \le \delta(E) \le \liminf_{k \to +\infty} \frac{\log N_{\frac{1}{a_{2k}}}(E)}{\log a_{2k}}.$$

Henceforth,

$$\dim E \le \varliminf_k \frac{\displaystyle\sum_{j=1}^{k} \log b_{2j-1}}{\displaystyle\sum_{j=1}^{2k} \log b_j}.$$

Taking $b_j = 2^{2^{j!}}$, we obtain

$$\dim E = \dim F = 0.$$

4.7 Exercises for Chapter 4

Exercise 1.
Consider the collection $\mathcal{F}$ composed Balls $B(x,r)$ in $\mathbb{R}^d$. Show that $\mathcal{F}$ permits the computation of the Hausdorff dimension of sets in $\mathbb{R}^d$.
Hint. We may prove firstly that, for all $\alpha \ge 0$, we have

$$\mathcal{H}^\alpha(E,\mathcal{F}) \le 2^\alpha \mathcal{H}^\alpha(E) \le 2^\alpha \mathcal{H}^\alpha(E,\mathcal{F}).$$

Exercise 2.
Show that the collection $\mathcal{F}$ of dyadic intervals in $\mathbb{R}$ permits the computation of the Hausdorff dimension. Recall that, for an integer $c \ge 2$, we call c-c-adic interval of order $n \in \mathbb{N}$ any interval of the form

$$I_p^n(c) = [\frac{p}{c^n}, \frac{p+1}{c^n}[, \quad p \in \mathbb{Z}.$$

Hint. We may prove as in Exercise 1 that, for all $\alpha \ge 0$, we have

$$\mathcal{H}^\alpha(E,\mathcal{F}) \le 3\mathcal{H}^\alpha(E) \le 3\mathcal{H}^\alpha(E,\mathcal{F}).$$

Exercise 3.
Let $F \subset \mathbb{R}^d$, and $\Phi : F \to \mathbb{R}^d$ be such that,

$$C_2|x - y| \le |\Phi(x) - \Phi(y)| \le C_1|x - y|, \quad \forall x,y \in F,$$

($C_1, C_2 > 0$ constants). Show that $\dim \Phi(F) = \dim F$.

Exercise 4.
Let $F \subset \mathbb{R}^d$ be such that, $\dim F < 1$. Show that F is totally discontinuous: Each connected component of F is reduced to one point.

Hint. We may consider, for $x \in F$ fixed, the function $f : \mathbb{R}^d \to [0, +\infty[$, such that, $f(z) = |z - x|$, and prove next, for $x \neq y$ in F, there exists Θ_x, Θ_y two open subsets of F, such that,

$$\Theta_x \cap \Theta_y = \phi, \quad x \in \Theta_x, y \in \Theta_y, \quad \text{and} \quad \Theta_x \bigcup \Theta_y = F.$$

Exercise 5.

Let K_1, and K_2 be compact subsets in $\mathbb{R}^{d_1}$, and $\mathbb{R}^{d_2}$, respectively. d_1, d_2 are integers. Then,

$$\dim(K_1 \times K_2) \geq \dim K_1 + \dim K_2.$$

The equality is not true in general.

Exercise 6.

(1) Let $E \subset \mathbb{R}^n$, and μ be a Borel measure, such that, $\mu(E) > 0$, and that

$$\mu(B(x, r)) \leq C r^\alpha,$$

for some constant $0 < C < \infty$, and for any Ball $B(x, r)$, $x \in E$. Show that

$$H^\alpha(E) \geq H^\alpha(E) \geq \mu(E)/c,$$

and deduce that $dim(E) \geq \alpha$.

(2) Let μ be a probability measure on $A \subset [0, 1]$, satisfying

$$\mu(I) \leq C|I|^\alpha,$$

for all interval $I \subset [0, 1]$. Let next $F \subset [0, 1]^2$ be such that, $dim F \geq \alpha$, and denote

$$F_x = \{t \in [0, 1]; \ (x, t) \in F\}.$$

Show that

$$dim(F_x) \leq dim(F) - \alpha,$$

for μ-almost every x in A.

Exercise 7.

Let $0 \leq \alpha \leq \frac{1}{2}$.

1. Plot the set $\mathcal{C}$ of points in the complex plane as follows:

- Start with the line segment $[0, 1]$
- At each stage replacing each interval $I = [x, y]$ by the union of intervals

$$[x, z] \cup [z, w] \cup [w, z] \cup [z, y],$$

where

$$z = \frac{1}{2}(x + y), \quad \text{and} \quad w = z + i\alpha(y - x).$$

2. Compute the Hausdorff dimension of the resulting set.

3. Describe the exact nature of the set $\mathcal{P}$ for $\alpha = \frac{1}{2}$.

4. Consider next the Hausdorff distance defined for two compact sets A, B as

$$d_H(A, B) = \max\{\max_{a \in A} dist(a, B), \max_{b \in B} dist(b, A)\}.$$

If $A_n \to A$ in the Hausdorff metric, does $dim(A_n) \to dim(A)$?

Exercise 8.

Let E be a Cantor set in $\mathbf{R}^n$. Is the projection of E onto a k-dimensional subspace is necessarily a Cantor set in $\mathbf{R}^k$?

Exercise 9.

Let $K \subset \mathbf{R}^n$ be the Cantor set associated to the similitudes $\{f_1, \ldots, f_n\}$, that satisfy the open set condition. Prove that the set associated to $\{f_1, \ldots, f_{n-1}\}$ is a subset of strictly smaller dimension.

Exercise 10.

Let $A \subset \mathbf{R}^d$, and $B \subset \mathbf{R}^n$ be compact sets.

a. Show that

$$dim(A) + dim(B) \leq dim(A \times B) \leq dim(A) + dim(B).$$

b. Give an example of subsets E, and F of $\mathbf{R}$ such that

$$dim(E) + dim(F) \leq dim(E \times F) < dim(E) + dim(F) + 1.$$

Chapter 5

Capacity Dimension of Sets

As it is mentioned in the previous chapters, the main problem in computing the Hausdorff dimension of sets appeared directly from its mathematical definition as an outer measure, by means of coverings. This is why researchers in different fields have proposed many variants to estimate the Hausdorff dimension, and/or its modified variants, especially when dealing with applications. One of the well known variants of Hausdorff dimension is the so-called Capacity dimension of sets, which will be studied in the present chapter. Such a dimension has many advantages, and applications. It relates, for example, the theory of Hausdorff, and generally fractal dimensions to physics, as it is strongly related to the physical Capacity of physical instruments such as conductors.

For detailed study, and more backgrounds on the concept of Capacity as well as its relation, and application in different fields, the readers may refer to [Beardon (1965); Bélair (1987); Billingsley (1960, 1961, 1965); Boyd (1973); Buck (1970, 1973); Choquet (1961); David and Semmes (1997); Edgar (1998, 2008); Eggleston (1949, 1951); Falconer (1994, 1990); Frostman (1935); Kigami (2001); Pesin (1997); Selezneff (2011)]

5.1 Generalities

Let μ be a probability measure on $\mathbb{R}^d$, and $\alpha > 0$. Denote

$$I_\alpha(\mu) = \int \int \frac{d\mu(x)d\mu(y)}{|x - y|^\alpha},$$

where, for a vector $x \in \mathbb{R}^d$, the notation $|x|$ designates its Euclidean norm. We will denote also for a subset A in $\mathbb{R}^d$ its diameter by $|A|$ or $diam(A)$, which is evaluated as

$$|A| = diam(A) = \sup_{x,y \in A} |x - y|.$$

Definition 5.1. Let A be a compact subset in $\mathbb{R}^d$. A is said to be an α-positive Capacity, and we write $Cap_\alpha A > 0$, if there exists a probability measure μ on $\mathbb{R}^d$ supported by A, such that, $I_\alpha(\mu) < \infty$. Otherwise, we write $Cap_\alpha A = 0$.

We immediately have the following characterization.

Lemma 5.1. $Cap_\alpha A = 0 \;\Rightarrow\; Cap_\gamma A = 0, \; \forall \gamma > \alpha$.

Proof. We have

$$I_\gamma(\mu) = \int \int \frac{d\mu(x)d\mu(y)}{|x-y|^\gamma} = \int \int \frac{d\mu(x)d\mu(y)}{|x-y|^{\gamma-\alpha}|x-y|^\alpha} \geq \frac{I_\alpha(\mu)}{|A|^{\gamma-\alpha}}.$$

As $Cap_\alpha A = 0$, we get $I_\alpha(\mu) = \infty$. Hence, $I_\gamma(\mu) = \infty$. So, $Cap_\gamma A = 0$. $\square$

Definition 5.2. The Capacity dimension of A is defined by

$$\dim_c A = \inf\{\alpha > 0, \; Cap_\alpha = 0\}.$$

Theorem 5.1. *Let A be a compact in $\mathbb{R}^d$, then,*

$$\dim_c A = \dim A.$$

Proof. Let $0 < \alpha < \gamma$, and μ be a probability measure on $\mathbb{R}^d$, such that, $\mu(A) = 1$. There exists $R > 0$, such that, $|x-y| \leq R, \; \forall x, y \in A$. Consequently,

$$I_\gamma(\mu) = \int \int \frac{d\mu(y)d\mu(x)}{|x-y|^\gamma} \geq \frac{1}{R^{\gamma-\alpha}} \int \left(\int \frac{d\mu(y)}{|x-y|^\alpha} \right) d\mu(x) \geq \frac{I_\alpha(\mu)}{R^{\gamma-\alpha}}.$$

Therefore, it suffices to show that

$$\mathcal{H}^\gamma(A) > 0 \;\Rightarrow\; Cap_\alpha A > 0 \;\Rightarrow\; \mathcal{H}^\alpha(A) > 0.$$

Indeed, it is straightforward that

$$\dim A = 0 \;\Longrightarrow\; Cap_s A = 0, \; \forall s > 0.$$

Consequently, $\dim_c A < s, \; \forall s > 0$, which means that $\dim_c A = 0$. Next, for $\dim A > 0$, let γ be such that, $0 < \gamma < \dim A$. We immediately obtain

$$Cap_\gamma A > 0 \Longrightarrow \dim_c A \geq \gamma.$$

Consequently,

$$\dim_c A \geq \dim A.$$

Now, let α be such that, $0 < \alpha < \dim_c A$. We have

$$I_\alpha(\mu) = \infty \;\Longrightarrow\; \dim A \geq \alpha \;\Longrightarrow\; \dim A \geq \dim_c A.$$

Indeed, there exists a probability measure μ on $\mathbb{R}^d$ such that $\mu(A) = 1$, and
$$\mu(B) \leq M|B|^\gamma.$$
Consequently,
$$\mu(\{x\}) = 0, \quad \forall\, x \in A.$$
Now, observe that
$$\int \frac{d\mu(y)}{|x - y|^\alpha} = \int_{0 < |y-x| < 1} \frac{d\mu(y)}{|x - y|^\alpha} + \int_{|y-x| \geq 1} \frac{d\mu(y)}{|x - y|^\alpha}.$$
Denote
$$I = \int_{|y-x| \geq 1} \frac{d\mu(y)}{|x - y|^\alpha},$$
and
$$J = \int_{0 < |y-x| < 1} \frac{d\mu(y)}{|x - y|^\alpha}.$$
We have
$$J = \sum_{n \geq 1} \int_{\frac{1}{2^n} \leq |x-y| < \frac{1}{2^{n-1}}} \frac{d\mu(y)}{|x - y|^\alpha}$$
$$\leq \sum_{n \geq 1} 2^{n\alpha} \mu(B((x, \tfrac{1}{2^{n-1}})))$$
$$\leq M \sum_{n \geq 1} 2^{n(\gamma - \alpha)} < \infty.$$
Hence, $I_\alpha(\mu) < \infty$. Conversely, let $\alpha > 0$ be such that, $Cap_\alpha A > 0$. We shall show that $\mathcal{H}^\alpha(A) > 0$. Indeed, there exists a Borel probability measure μ on $\mathbb{R}^d$, such that, $\mu(A) = 1$, and $I_\alpha(\mu) < \infty$. This yields that there exists $t > 0$, such that, $\mu(A_t) > 0$, where
$$A_t = \Big\{x \in A, \int \frac{d\mu(y)}{|x - y|^\alpha} \leq t\Big\}.$$
Whenever $A_t \subseteq \bigcup_j B_j$, with $B_j \cap A_t \neq \emptyset$, we get, for $x \in A_t \cup B_j$,
$$|x - y| \leq |B_j|, \quad \forall\, y \in B_j,$$
and thus,
$$\frac{\mu(B_j)}{|B_j|^\alpha} \leq \int_{B_j} \frac{d\mu(y)}{|x - y|^\alpha} \leq t.$$
As a consequence,
$$\frac{1}{t}\mu(B_j) \leq |B_j|^\alpha,$$
which yields that
$$0 < \frac{1}{t}\mu(A_t) \leq \sum |B_j|^\alpha.$$
As a result, $\mathcal{H}^\alpha(A_t) > 0.$ $\qquad\qquad\square$

5.2 Self-similar sets

Definition 5.3. A function $f : \mathbb{R}^d \longrightarrow \mathbb{R}^d$ is said to be contractive or a Contraction of $\mathbb{R}^d$ if there exists a constant c, $0 < c < 1$, such that,

$$|f(x) - f(y)| \leq c|x - y|; \ \forall x, y \in \mathbb{R}^d.$$

In the case of equality we say that f is a self-similar function or a self-similar with ratio c.

Consider next a finite set of contractions $S = (S_i)_{1 \leq m}$ on $\mathbb{R}^d$. We define by induction a set of set-valued maps $S^k(E)$, $k \geq 0$ as follows. For $E \subset \mathbb{R}^d$, let

$$S^0(E) = E, \ S^1(E) = S(E) = \bigcup_{i=1}^{n} S_i(E),$$

and

$$S^{k+1}(E) = S(S^k(E)), \ \forall k.$$

Definition 5.4. A subset $F \subset \mathbb{R}^d$ is said to be S-invariant, if it satisfies

$$S(F) = \bigcup_{i=1}^{n} S_i(F) = F.$$

The following result deals with the existence, and uniqueness of S-invariant sets .

Theorem 5.2. *Let $S = (S_i)_{1 \leq i \leq m}$ be a finite set of contractions on $\mathbb{R}^d$. There exists a unique non-empty compact F in $\mathbb{R}^d$ which is S-invariant. Furthermore,*

$$F = \bigcap_{k \geq 0} S^k(A),$$

for all non-empty compact A in $\mathbb{R}^d$, such that, $S(A) \subset A$.

Proof. We will apply the fixed point theorem. Let $\mathcal{C}$ be the collection of all compact subsets in $\mathbb{R}^d$, and consider the map $S : \mathcal{C} \longrightarrow \mathcal{C}$, which maps a compact subset X to its image $S(X)$ defined by $S(X) = \bigcup_{i=1}^{n} S_i(X)$. We shall consider a distance δ on $\mathcal{C}$, for which,

- $(\mathcal{C}, \delta)$ is a complete metric space.
- S is contractive on $\mathcal{C}$:

$$S(X), S(Y)) \leq c_\delta \delta(X, Y), \ \forall X, Y \in \mathcal{C},$$

and for some constant c_δ, $0 < c_\delta < 1$.

To do it consider the Hausdorff distance δ defined as follows. For $X \in \mathcal{C}$, and $a > 0$, denote

$$B(X, a) = \Big\{ y \in \mathbb{R}^d; \ d(y, X) \leq a \Big\} = \bigcup_{x \in X} B'(x, a).$$

It is straightforward that $B(X, a)$ is a compact subset of $\mathbb{R}^d$, and thus, $B(X, a) \in \mathcal{C}$. The distance δ will be defined by

$$\delta(X, Y) = \inf \Big\{ a > 0; \ X \subset B(Y, a), \text{ and } Y \subset B(X, a) \Big\}.$$

Next, we shall show that $(\mathcal{C}, \delta)$ is complete. Let $(X_n)_{n \geq 1}$ be a Cauchy sequence in $(\mathcal{C}, \delta)$. For all $\epsilon > 0$, there exists $p \in \mathbb{N}$, such that,

$$\delta(X_n, X_m) < \epsilon, \ \forall m \geq n \geq p.$$

Hence, the set $\bigcup_{i=1}^{\infty} X_i$ is bounded, due to the fact that

$$X_n \subset B(X_p, \epsilon), \ \forall n \geq p.$$

Denote next

$$X = \bigcap_i \overline{\Big(\bigcup_{j \geq i} X_j \Big)}.$$

It is straightforward that X is a non-empty compact subset of $\mathbb{R}^d$, and thus, $X \in \mathcal{C}$. Moreover,

$$\lim_{n \hookrightarrow +\infty} X_n = X.$$

Indeed, for $\epsilon > 0$, and p defined previously, we have

$$X_j \subset B(X_n, \epsilon), \ \forall j \geq n \geq p.$$

As a result,

$$\overline{\bigcup_{j \geq i} X_j} \subset B'(X_n, \epsilon), \ \forall j \geq n \geq p.$$

Consequently,

$$X \subset B(X_n, \epsilon).$$

Conversely, let $x \in X_n$, $n \geq p$. We immediately observe that

$$X_n \subset B(X_k, \epsilon), \ \forall k \geq n \geq p.$$

Therefore, for $k \geq n \geq p$, there exists $y_k \in X_k$, such that, $d(x, y_k) \leq \epsilon$. Observe also that $y_k \in \overline{\bigcup_{j \geq k} X_j}$. There exists consequently a sub-sequence

$(y_{n_k})_k$ convergent to y in X, and satisfying, $d(x, y_{n_k}) \leq \epsilon$, which means that $d(x, y) \leq \epsilon$. So, $d(x, X) \leq \epsilon$, which yields that

$$x \in B(X, \epsilon),$$

and thus,

$$X_n \subset B(X, \epsilon).$$

Consequently,

$$\delta(X_n, X) \leq \epsilon, \ \forall n \geq p.$$

As a result, we conclude that (φ, δ) is complete. It remains to show that S is a Contraction relatively to the distance δ. We will show precisely that

$$\delta(S(X), S(Y)) \leq \sup_{1 \leq i \leq m} \delta(S_i(X), S_i(Y)). \tag{5.1}$$

Let $a > \sup_{1 \leq i \leq m} \delta(S_i(X), S_i(Y))$. Then, $\forall i, 1 \leq i \leq m :$, we have

$$S_i(X) \subset B(\bigcup_{j=1}^{m} S_j(Y), a) \ , \text{ and } \ S_i(Y) \subset B(\bigcup_{j=1}^{m} S_j(X), a).$$

Consequently,

$$\bigcup_{1 \leq i \leq m} S_i(X) \subset B(\bigcup_{j=1}^{m} S_j(Y), a),$$

and

$$\bigcup_{1 \leq i \leq m} S_i(Y) \subset B(\bigcup_{j=1}^{m} S_j(X), a).$$

Hence,

$$\delta\left(\bigcup_{1 \leq i \leq m} S_i(X), \bigcup_{1 \leq i \leq m} S_i(Y) \right) \leq a.$$

This is true

$$\forall a > \sup_{1 \leq i \leq m} \delta(S_i(X), S_i(Y)).$$

So, equation (5.1) holds. Now, as we already have

$$\delta(S_i(X), S_i(Y)) \leq c_i \delta(X, Y), \ \forall i,$$

where c_i is the ratio of the Contraction S_i, we take $C_\delta = \sup\limits_{1 \leq i \leq m} c_i$. We thus, conclude that S is contractive on $\mathcal{C}$. So, finally, there exists a unique compact subset F of $\mathbb{R}^d$, such that,

$$S(F) = F = \bigcup_{i=1}^{m} S_i(F).$$

It remains now to prove the last point in the theorem. To do it, we will show some more general result. Let $(X_k)_k$ be a decreasing sequence in $\mathcal{C}$, and denote

$$X = \lim_{k \to \infty} X_k = \bigcap_k X_k.$$

We claim that $\delta(X_k, X)$ is decreasing to 0 in $\mathbb{R}$. Indeed, the sequence $(\delta(X_k, X))_k$ is decreasing. If it is bounded away from 0, there exists $a > 0$, such that,

$$\delta(X_k, X) \geq a > 0, \ \forall k.$$

This yields that

$$X_k \not\subseteq B(X, a), \ \forall k.$$

Hence, $\forall k$, there exists $x_k \ (\in X_k)$, such that,

$$d(x_k, X) \geq a.$$

On the other hand, there exists a sub-sequence $(x_{n_k})_k$ convergent to some point $x \in X$. Consequently,

$$\lim_{n \to \infty} d(x_{n_k}, X) = d(x, X) = 0 \geq a > 0,$$

which is contradictory. As a result,

$$\delta(X_k, X) \to 0, \ \text{as} \ k \to \infty,$$

which means that $(X_k)_k$ converges to X in $(\mathcal{C}, \delta)$. Now, for $X_k = S^k(A)$, we get a decreasing sequence, which by the previous result, converges to

$$X = \bigcap_k X_k = \bigcap_k S^k(A).$$

Hence, $F = \bigcap_{k \geq 0} S^k(A)$. $\qquad\qquad\qquad\qquad\qquad\qquad\qquad\qquad\qquad$ $\square$

Theorem 5.3. *Let $S = (S_i)_{1 \le i \le m}$ be a set of similarities with respective ratios $(c_i)_{1 \le i \le m}$. Let F be the unique non-empty compact S-invariant. Assume that S satisfies the open set condition: There exists a non-empty open, and bounded set V in $\mathbb{R}^d$, such that*

$$S(V) = \bigcup_{i=1}^{m} S_i(V) \subset V \quad and \quad S_i(V) \bigcap S_j(V) = \phi, \ \forall i \ne j. \qquad (5.2)$$

Then, $\dim F = s$, where s is the unique solution of $\sum_{i=1}^{n} c_i^s = 1$. Besides, we have

$$0 < \mathcal{H}^s(F) < \infty.$$

Example 5.1. We propose to compute the Hausdorff, and the Capacity dimension of the triadic Cantor set, by applying Theorem 5.3. We consider the set $S = (S_1, S_2)$, where the contractions S_1, and S_2 are defined on $\mathbb{R}$ by

$$S_1(x) = \frac{1}{3}x \quad and \quad S_2(x) = \frac{1}{3}x + \frac{2}{3}.$$

Let F be the unique non-empty compact S-invariant. We claim that $F = C$, the triadic Cantor set. To do this, consider $A = [0, 1]$. We have

$$S(A) = [0, \frac{1}{3}] \cup [\frac{2}{3}, 1] \subset A.$$

Consequently,

$$F = \bigcap_{k \ge 1} S^k(A).$$

It remains to show that

$$S^k(A) = E_k, \ \forall k.$$

For $k = 0$, we have

$$S^0(A) = A = [0, 1] = E_0.$$

For $k = 1$,

$$S^1(A) = [0, \frac{1}{3}] \bigcup [\frac{2}{3}, 1] = E_1.$$

So, whenever for a step k, $S^k(A) = E_k$, we get

$$
\begin{aligned}
S^k(A) &= E_k \\
&= \bigcup_{(i_1,\dots,i_k) \in \{1,2\}^k} S_{i_1} \circ \cdots \circ S_{i_k}(A) \\
&= \bigcup_{(i_1,\dots,i_k) \in \{1,2\}^k} I_{i_1,\dots,i_k},
\end{aligned}
$$

where $I_{i_1,\ldots,i_k}$ are the triadic intervals composing E_k. This yields that

$$S^{k+1}(A) = \bigcup_{(i_1,\ldots,i_k)\in\{1,2\}^k} \Big(S_1(I_{i_1,\ldots,i_k}) \cup S_2(I_{i_1,\ldots,i_k})\Big).$$

Otherwise,

$$S^{k+1}(A) = \bigcup_{(i_1,\ldots,i_k,i_{k+1})\in\{1,2\}^k} I_{i_1,\ldots,i_k,i_{k+1}} = E_{k+1}.$$

Finally, to compute the Hausdorff dimension, we chose $V =]0,1[$. We immediately have

$$S(V) \subset V, \text{ and } S_1(V)\bigcap S_2(V) = \phi.$$

So, $\dim C = s$, where s is the unique solution of

$$c_1^s + c_2^s = 1,$$

or equivalently,

$$2\Big(\frac{1}{3}\Big)^s = 1,$$

which yields that

$$s = \frac{\log 2}{\log 3}.$$

Theorem 5.4. *Let $S = (S_i)_{1\leq i\leq m}$ be a set of contractive self-similarities, with respective ratios $(c_i)_{1\leq i\leq m}$. Assume further that S satisfies the open set condition (5.2). Then, there exists a Borel probability measure μ on $\mathbb{R}^d$, supported by F, such that, $\forall X$ a Borel subset of $\mathbb{R}^d$, we have*

$$\mu(X) = \sum_{i=1}^{m} c_i^s \mu(S_i^{-1}(X)),$$

where s is the unique solution of

$$\sum_{i=1}^{m} c_i^s = 1.$$

Proof. Let $x \in F$ be fixed, and denote for $k \in \mathbb{N}$, the linear form $L_k : C_c(\mathbb{R}^d) \to \mathbb{R}$, such that,

$$L_k(f) = \sum_{(i_1,\ldots,i_k)\in\{1,\ldots,m\}^k} c_{i_1}^s \ldots c_{i_k}^s f(x_{i_1,\ldots,i_k}),$$

where $x_{i_1}...i_k = S_{i_1} \circ ...S_{i_k}(x)$. It holds that $(L_k(f))_k$ is a Cauchy sequence in $\mathbb{R}$. Let L be the linear form limit. Riesz theorem yields that there exists a Borel probability measure μ on $\mathbb{R}^d$, such that,

$$L(f) = \int f d\mu.$$

Standard techniques permit to conclude that

$$\mu(\mathbb{R}^d) = \mu(F) = 1.$$

On the other hand, we have

$$\int f d\mu = \sum_{i=1}^{m} c_i^s \int f \circ S_i d\mu, \ \forall f \in C_c(\mathbb{R}^d).$$

Consequently, by the density of $C_c(\mathbb{R}^d)$ in the L^p spaces, we get

$$\int f d\mu = \sum_{i=1}^{m} c_i^s \int f \circ S_i d\mu,$$

for any integrable function f. So, for $f \equiv 1_X$, we obtain

$$\mu(X) = \sum_{i=1}^{m} c_i^s \mu(S_i^{-1}(X)).$$

$\square$

Lemma 5.2. *Let $(V_i)_{i \in I}$ be a family of disjoint open sets in $\mathbb{R}^d$. Let $0 < r_1 < r_2$, $\rho > 0$, be such that, $\forall i$, V_i contains a Ball of radius $r_1\rho$, and is contained in a Ball of radius $r_2\rho$. Then, every Ball of radius ρ, intercepts, at most $(1 + 2r_2)^d r_1^{-d}$ elements of the family $(\overline{V_i})_{i \in I}$.*

Proof of Theorem 5.3. It suffices to prove that

$$0 < H^s(F) < +\infty.$$

Assume that

$$J_k = \{(i_1, ..., i_k), \ 1 \le i_j \le m\}.$$

For $A \subset \mathbb{R}^d$, let

$$A_{i_1,...,i_k} = S_{i_1} \circ ... \circ S_{i_k}(A).$$

We have

$$F = \bigcup_{i=1}^{m} S_i(F) = \bigcup_{(i_1,...,i_k) \in J_k} F_{i_1...i_k}.$$

Moreover,

$$|F_{i_1...i_k}| = c_{i_1}...c_{i_k}|F|.$$

So that,

$$\sum_i |F_{i_1...i_k}|^s = \sum_i c_{i_1}^s...c_{i_k}^s|F|^s = |F|^s.$$

Hence, $H^s(F) < +\infty$. Theorem 5.4 implies that there exists a Borel probability measure μ on $\mathbb{R}^d$, carried by F, such that, for any Borel set X in $\mathbb{R}^d$, we have

$$\mu(X) = \sum_{i=1}^{m} c_i^s \mu(S_i^{-1})(X).$$

We claim that μ is s-Hölderian. Indeed, let $0 < \rho < \min_i c_i$. We shall prove that, for any Ball $B(.,\rho)$, we have

$$\mu(B(.,\rho)) \leq C|B(.,\rho)|^s,$$

where $C > 0$ is a constant. Denote next

$$\mu_{i_1...i_k}(X) = \mu((S_{i_1} \circ ... \circ S_{i_k})^{-1}(X)).$$

Hence, $\mu_{i_1...i_k}$ is a probability measure on $\mathbb{R}^d$. S_i being bijective, so

$$\mu_{i_1...i_k}(F_{i_1...i_k}) = 1, \text{ and } F_{i_1...i_k} \subset \overline{V}_{i_1...i_k}. \tag{5.3}$$

Now, observe that

$$\mu = \sum c_i^s \mu_i = ... = \sum c_{i_1}^s...c_{i_k}^s \mu_{i_1...i_k},$$

and let $\eta_1 < \eta_2$ be such that, $B(.,\eta_1) \subset V \subset B(.,\eta_2)$. Then,

$$B(.,c_{i_1},...,c_{i_k}\eta_1) \subset V_{i_1...i_k} \subset B(.,c_{i_1},...,c_{i_k}\eta_2).$$

Let next $k_0 \in \mathbb{N}$ be such that,

$$(\max c_i)^{k_0} < \rho \min c_i,$$

and let k be the smallest integer, for which,

$$(i_1,...,i_{k_0}) \in J_{k_0}, \; 1 < k < k_0,$$

and

$$(\min c_i)\rho \leq c_{i_1}...c_{i_k} \leq \rho.$$

We define the mapping $\Phi : J_{k_0} \to \Phi(J_{k_0})$, by

$$\Phi((i_1,...,i_{k_0})) = (i_1,...,i_k).$$

We get

$$\mu = \sum_{(i_1...i_{k_0}) \in J_{k_0}} c_{i_1}^s ... c_{i_k}^s \mu_{i_1,...,i_{k_0}}$$

$$= \sum_{(j_1...j_k) \in \Phi(J_{k_0})} \sum_{(i_1,...,i_{k_0}) \in J_0} c_{i_1}^s ... c_{i_{k_0}}^s \mu_{i_1...i_{k_0}}.$$

Observing that

$$\Phi^{-1}(j_1,..,j_k) = \{(j_1,..,j_k,l_{k+1},..,l_{k_0})/l_j \in \{1,...,m\}\},$$

we obtain

$$\mu = \sum_{(j_1,...,j_k)} c_{j_1}^s ... c_{j_k}^s \sum_{l_j} c_{l_{k+1}}^s ... c_{l_{k_0}}^s \mu_{i_{k+1}...l_{k_0}}$$

$$= \sum c_{j_1}^s ... c_{j_k}^s \mu_{j_1...j_k}.$$

Otherwise,

$$(\min c_i)\rho \le c_{j_1}...c_{j_k} \le \rho.$$

So that,

$$(\min c_i)\eta_i \rho \le c_{j_1}...c_{j_k}\eta_1.$$

As a result,

$$B(.,(\min c_i)\eta_1\rho) \subset B(.,c_{j_1}...c_{j_k}\eta_1)$$
$$\subset V_{j_1...j_k}$$
$$\subset B(.,c_{j_1}...c_{j_k}\eta_2)$$
$$\subset B(.,\eta_2\rho).$$

On the other hand,

$$(\min c_i)\rho \le c_{j_1}...c_{j_k} \le \rho,$$

which yields that

$$(\min c_i)\eta_1\rho \le c_{j_1}...c_{j_k}\eta_1.$$

Hence,

$$B(.,(\min c_i)\eta_1\rho) \subset B(.,c_{j_1}...c_{j_k}\eta_1)$$
$$\subset V_{j_1...j_k}$$
$$\subset B(.,c_{j_1}...c_{j_k}\eta_2)$$
$$\subset B(.,\eta_2\rho).$$

Then, Lemma 5.2, with $r_1 = (\min c_i)\eta_1$, the family $(V_{j_1...j_k})$, and $r_2 = \eta_2$, yields that, $B(.,\rho)$ meets at most $N = (1+2r_2)^d r_1^{-d}$ elements of the family $(\overline{V}_{j_1...j_k})$. Thus,

$$\mu(B(.,\rho)) = \sum_{(j_1,...,j_k)} c_{j_1}^s ... c_{j_k}^s \mu_{(j_1,...,j_k)}(B(.,\rho))$$

$$\le N\rho^s$$

$$= \frac{N}{2^s}|B(.,\rho)|^s.$$

So, μ is s-Hölderian. $\qquad\qquad\square$

Proof of Lemma 5.2. Whenever, we have

$$R \geq \rho(X + 2r_2),$$

and

$$NC_d(r_1\rho)^d \leq C_d(1 + 2r_2)^d\rho^d,$$

we immediately deduce that

$$N \leq (1 + 2r_2)^d r_1^{-d}.$$

$\square$

Proof of Lemma 5.4. Let $x \in F$ be fixed. For $k \in \mathbb{N}$, consider the linear form L_k defined previously, and fix $f \in C_c(\mathbb{R}^d)$. We have, $\forall k, p \in \mathbb{N}$,

$$L_k(f) - L_p(f) = \sum (c_{i_1}...c_{i_p})^s \left(f(x_{i_1...i_k}) - f(x_{i_1...i_p}) \right).$$

Now, observing that

$$(f(x_{i_1...i_k}), f(x_{i_1...i_p})) \in F_{i_1...i_k}^2.$$

We obtain

$$|F_{i_1...i_k}| = c_{i_1}...c_{i_k}|F|.$$

Next, as f is uniformly continuous, then, $(L_k(f))_k$ is a Cauchy sequence in $\mathbb{R}$. Denote

$$L(f) = \lim_{k \to +\infty} L_k(f).$$

It follows that L is a linear positive form. Therefore, there exists a Borel measure $\mu \geq 0$ on $\mathfrak{M}$, a σ-algebra on $\mathbb{R}^d$, such that,

$$L(f) = \int_{\mathbb{R}^d} f d\mu \quad f \in C_c(\mathbb{R}^d).$$

In particular, there exists $(f_n)_n \subset C_c(\mathbb{R}^d)$ Increasing to $f = \chi_{\mathbb{R}^d}$, and for n large enough, $F \subset B(0, n)$. On the other hand, using $L_k(f_n)$, we get for $n \gg 1$,

$$\lim_{k \to +\infty} \int f_n d\mu = 1,$$

and

$$\int f d\mu = \mu((\mathbb{R}^d)) = 1.$$

We now claim that

$$\mu(F) = 1 \ (\Longleftrightarrow \ \mu(F^c) = 0).$$

Indeed, let $f \equiv 1$ on K, $support(f) \subset F_c$, and $f \in C_c((\mathbb{R}^d))$. We get

$$\mu(K) \leq \int f d\mu = \lim_{k \to +\infty} L_k(f) = 0, \quad \forall K \subset F^c.$$

It results that

$$\sup_{K \subset F^c} \mu(K) = \mu(F^c) = 0 \Rightarrow \mu(F) = 1.$$

Next, observing that, for $k \geq 2$,

$$f(x_{i_1 \ldots i_k}) = f \circ S_{i_1}(x_{i_2 \ldots i_k}),$$

we get

$$L_k(f) = \sum_{i=1}^{m} c_i^s L_{k-1}(f \circ S_i).$$

Letting $k \to \infty$, we obtain

$$\int f d\mu = \sum c_i^s \int f \circ S_i d\mu, \quad \forall f \in C_c(\mathbb{R}^d).$$

Using the density of $C_c(\mathbb{R}^d)$, we obtain the desired result. $\qquad \square$

Proposition 5.1. *Let $N \in \mathfrak{M}$ be such that, $\mu(N) = 0$, then*

$$S_j^{-1}(N) \in \mathfrak{M} , \quad and \quad \mu(S_j^{-1}(N)) = 0.$$

Proof. Let $f_n \to 1_X$, and $\mu(N) = 0$. Hence, $f_n \circ S_i \longrightarrow 1_X \circ S_i$, μ almost everywhere. Now, notice that $|f_n \circ S_i| \leq 1$, and

$$S_i^{-1}(C_{\mathbb{R}} N_d) = C_{\mathbb{R}} S_i^{-1}(N).$$

Let V be a bounded open set, with

$$\mu(S_j^{-1}(V)) \leq \frac{1}{c_j^s} \mu(V),$$

and consider

$$V_n = \{x \in V, d(x, \partial V) > \frac{1}{n}\}.$$

So, V_n is open, and $\overline{V_n}$ is compact $\subset V$. Let next $f \in C_c(\mathbb{R}^d)$ be such that, $0 \leq f \leq 1$, $f \equiv 1$ on V_n, and $support(f) \subset V$. We get

$$c_j^s \mu(S_j^{-1}(V_n)) \leq c_j^s \int f \circ S_j d\mu \leq \int f d\mu = \mu(V).$$

Now, remark that

$$V_n \uparrow V = \bigcup_n V_n , \quad and \quad S_j^{-1}(V) = \bigcup_n S_j^{-1}(V_n).$$

So, for arbitrary open set V, we chose $V_n = V \cap B(0, n)$, and we use the result proved previously. Let next $N \in \mathfrak{M}$ be such that, $\mu(N) = 0$. There exists V_n open, such that, $N \subset V_n$, and $\mu(V_n) \leq \frac{1}{n}$. Denote $W_n = \bigcap_{k=1}^n V_k$. So, W_n is open,

$$W_n \downarrow W = \bigcap_{n=1}^{\infty} W_n,$$

and $\mu(W_n) \leq \frac{1}{n}$. We have

$$S_j^{-1}(N) \subset S_j^{-1}(W) = \bigcap_{n=1}^{\infty} S_j^{-1}(W_n) \in \mathbf{B} \subset \mathfrak{M},$$

and

$$\mu(S_j^{-1}(W_n)) \leq \frac{1}{c_j^s} \frac{1}{n}.$$

Consequently,

$$\mu(S_j^{-1}(W)) = 0,$$

which implies that

$$\mu(S_j^{-1}(N)) = 0.$$

$\square$

5.3 Billingsley dimension

Let $c \in \mathbb{N}$, $c \geq 2$, and $\mathfrak{F}$ be the family of c-adic intervals $[\frac{k}{c^n}, \frac{k+1}{c^n}[$, where $k, n \in \mathbb{N}$. Let next μ be a non-atomic probability measure on $\mathbb{R}^d$. For $E \subset \mathbb{R}^d$, $\alpha \geq 0$, and $\epsilon > 0$, denote

$$\mathcal{H}_{\mu,\epsilon}^{\alpha}(E) = \inf\left\{\sum_j \mu(I_j)^{\alpha}, E \subset \bigcup_j I_j, I_j \in \mathfrak{F}, \mu(I_j) \leq \epsilon\right\}.$$

As for the case of the Hausdorff measure, we obtain here an Increasing function in ϵ. So, denote

$$\mathcal{H}_{\mu}^{\alpha}(E) = \lim_{\epsilon \to 0} \mathcal{H}_{\mu,\epsilon}^{\alpha}(E).$$

We get here an outer metric measure on $\mathbb{R}^d$. Furthermore, there exists a cut-off value $D \in \overline{\mathbb{R}}$ such that

$$\mathcal{H}_{\mu}^{\alpha}(E) = \infty, \quad \text{for} \quad \alpha < D, \quad \text{and} \quad \mathcal{H}_{\mu}^{\alpha}(E) = 0, \quad \text{for} \quad \alpha > D.$$

We write $\dim_{\mu} E = D$, and call it the μ-dimension of E.

Proposition 5.2. *The following assertions are true.*

(1) $0 \leq dim_\mu E \leq 1$.

(2) $dim_\mu \{a\} = 0, \forall a \in \mathbb{R}$.

(3) $dim_\mu \bigcup_n E_n = \sup_n dim_\mu E_n$.

(4) Whenever E is a Borel set, and $\mu(E) > 0$, then, $dim_\mu E = 1$.

(5) For $E \subset F$, we have $dim_\mu E \leq dim_\mu F$.

Proof. (1) From its definition, we see that $dim_\mu E \geq 0$. Let next $s > 1$, and $\varepsilon > 0$. For any ε-covering $(I_j)_j \subset \mathcal{F}$ of E, we have

$$\mathcal{H}^s_{\mu,\varepsilon}(E) \leq \sum_j \mu(I_j) \leq \mu\left(\bigcup_j I_j\right) \leq 1.$$

Consequently,

$$\mathcal{H}^s_\mu(E) \leq 1, \forall s > 1,$$

which means that $dim_\mu E \leq s, \forall s > 1$. Hence, $dim_\mu E \leq 1$.

(2) For all $n \in \mathbb{N}$, the singleton $\{a\}$ is contained in one interval $I_k \in \mathcal{F}$, (for which $\frac{k}{c^n} \leq a < \frac{k+1}{c^n}$). As a result, for all $\varepsilon > 0$, and all n large enough ($n > -\frac{\log \varepsilon}{\log c}$), we get

$$\mathcal{H}^\alpha_{\mu,\varepsilon}(\{a\}) \leq \frac{1}{c^{n\alpha}},$$

which yields that

$$\mathcal{H}^\alpha_\mu(\{a\}) < \infty, \forall \alpha \geq 0.$$

Consequently,

$$dim_\mu(\{a\}) \leq \alpha, \forall \alpha \geq 0.$$

With assertion 1, we get $dim_\mu(\{a\}) = 0$.

(3) It is obvious that any covering of F is a covering of E. Therefore, $\forall \alpha > \dim_\mu F$, we get

$$\mathcal{H}^\alpha_{\mu,\varepsilon}(E) = \leq \mathcal{H}^\alpha_{\mu,\varepsilon}(F), \forall \varepsilon > 0.$$

As a result,

$$\mathcal{H}^\alpha_\mu(E) \leq \mathcal{H}^\alpha_\mu(E) = 0.$$

This yields that

$$dim_\mu E \leq \alpha, \forall \alpha > \dim_\mu F.$$

Consequently, $dim_\mu E \leq dim_\mu F$.

(4) Let $\alpha > \sup_p dim_\mu E_p$. It holds from that

$$0 \leq \mathcal{H}^\alpha_\mu(\bigcup_p E_p) \leq \sum_p \mathcal{H}^\alpha_\mu(E_p) = 0.$$

Consequently,

$$\mathcal{H}_\mu^\alpha(\bigcup_p E_p) = 0,$$

which yields that

$$dim_\mu(\bigcup_p E_p) \leq \alpha, \quad \forall \alpha > \sup_p dim_\mu E_p.$$

As a result,

$$dim_\mu(\bigcup_p E_p) \leq \sup_p dim_\mu E_p.$$

The converse follows from assertion 3.

(5) Let $\alpha < 1$. For all $\varepsilon > 0$, and $(I_j)_j \subset \mathcal{F}$ and ε-covering of E, we have

$$\mu(E) \leq \sum_j \mu(I_j) \leq \sum_j \mu(I_j)^\alpha.$$

Therefore,

$$0 < \mu(E) \leq \mathcal{H}_{\mu,\varepsilon}^\alpha(E).$$

Letting $\varepsilon \downarrow 0$, we obtain

$$\mathcal{H}_\mu^\alpha(E) > 0, \quad \forall \alpha < 1.$$

As a results, $dim_\mu E \geq 1$. By assertion 1, we deduce the equality desired. $\square$

Theorem 5.5. *For $t \in \mathbb{R}$, let $I_n(t)$ be the c-adic interval of order n, which contains t. Let μ, ν be two non-atomic probability measures on $\mathbb{R}$, and $\alpha \geq 0$. Then, for any subset*

$$E \subset \left\{ t \in \mathbb{R}, \quad \lim_{n \to +\infty} \frac{\log \mu(I_n(t))}{\log \nu(I_n(t))} = \alpha \right\},$$

we have

$$dim_\nu E = \alpha \, dim_\mu E.$$

Proof. Assume that $0 < \alpha < +\infty$. We shall show, in a first step, that for

$$E \subset \{ t \in \mathbb{R}, \underline{\lim}_n \frac{\log \mu(I_n(t))}{\log \nu(I_n(t))} \geq \alpha \},$$

we have

$$dim_\nu E \geq \alpha dim_\mu E.$$

So, assume that, $\forall I_j \in \mathfrak{F}$, such that, $I_j \bigcap E \neq \emptyset$, we have $\nu(I_j) \neq 0$. Let next $\beta > dim_\nu E$, and $\eta > \frac{1}{\alpha}$. We claim that

$$\dim_\mu E \leq \eta\beta.$$

Indeed, let

$$E_k = \{t \in E, \ \forall n, \ \nu(I_n(t)) \geq \frac{1}{k}, \ \text{ or } \ \mu(I_n(H)) \leq \nu(I_n(H))\}.$$

Hence, $(E_k)_k \nearrow E = \bigcup_k E_k$. Indeed, $t \in E$, there exists p, such that,

$$\frac{\log \mu(I_n(t))}{\log \nu(I_n(t))} \geq \frac{1}{\eta}, \ \forall n \geq p.$$

Let next k be such that, $\frac{1}{k} \leq \nu(I_p(t))$. We get $t \in E_k$. Therefore, it suffices to prove that

$$\dim_\mu E_m \leq \eta\beta, \ \forall m.$$

Consider a rectangle $(I_j)_j$ of E_m, such that, $\forall j$,

$$I_j \in \mathfrak{F}, \ \text{ and } \ \nu(I_J) \leq \epsilon < \frac{1}{k}.$$

It follows that

$$\mu(I_j)^\eta \leq \nu(I_j), \ \forall j.$$

Consequently,

$$\mathcal{H}^{\eta\beta}_{\mu, \epsilon^{\frac{1}{\eta}}}(E_m) \leq \sum_j \mu(I_j)^{\eta\beta} \leq \sum \nu(I_j)^\beta,$$

which yields that

$$\mathcal{H}^{\eta\beta}_{\mu, \epsilon^{\frac{1}{\eta}}}(E_m) \leq \mathcal{H}^{\beta}_{\nu, \epsilon}(E_m).$$

Letting $\epsilon \to 0$, we get

$$\mathcal{H}^{\eta\beta}_\mu(E_m) \leq \mathcal{H}^\beta_\nu(E_m) = 0.$$

As a conclusion,

$$\dim_\mu E_m \leq \eta\beta.$$

We now investigate the general case. Assume firstly that $\dim_\mu F_\mu = 0$. Observing that

$$E = (E \setminus F_\nu) \bigcup (E \cap F_\nu),$$

we get

$$\dim_\nu E = \dim_\nu E \setminus F_\nu.$$

On the other hand, the elements of $E \setminus F_\nu$ satisfy the hypothesis of the first case. Therefore,

$$\dim_\nu E \setminus F_\nu \geq \alpha \dim_\mu E \setminus F_\nu \geq \alpha \dim_\mu E \setminus F_\mu = \alpha \dim_\mu E$$

as $F_\nu \subset F_\mu$. Assume now that $\alpha = 0$, and let $\eta > 0$ be such that, $\dim_\nu E \leq \eta$. We notice immediately that

$$E \subset \left\{ t \in \mathbb{R}, \ \overline{\lim}^n \frac{\log \mu(I_n(t))}{\log \nu(I_n(t))} \leq \eta \right\}.$$

Whenever $I_j \cap E \neq \emptyset$ satisfies $\mu(I_j) \neq 0$, we get

$$E_k = \left\{ t \in \mathbb{R}, \ \mu(I_n(t)) \geq \frac{1}{k}, \ \text{or} \ \nu(I_n(t))^\eta \leq \mu(I_n(t)) \right\}.$$

So, the same techniques as above yield that $E = \bigcup_k E_k$, and consequently, $\dim_\nu E_k \leq \eta$. □

5.4 Eggleston theorem

Let $c \geq 2$ in $\mathbb{N}$, denote $N_c = \{0, 1, \cdots, c-1\}$, and N_c^* the set of finite words constructed with N_c as an alphabet. For $j \in N_c^*$, $j = j_1 j_2, \ldots, j_n$, with $j_k \in N_c$, we write $|j| = n$ the length of the word j. For $k \in N_c$, let $N_k(j)$ be the number of appearance of the letter k in j, and

$$I_j = \left[\sum_{k=1}^n \frac{j_k}{c^k}, \sum_{k=1}^n \frac{j_k}{c^k} + \frac{1}{c^n} \right[.$$

It is clear that I_j is a c-c-adic interval $\subset [0, 1[$. Finally, for $k \in N_c$, and $t \in [0, 1[$ let $\varphi_n^k(t) = N_k(j)$, where $t \in I_j$, $|j| = n$.

Theorem 5.6. *Let $p = (p_k)_{0 \leq k \leq c-1}$ be a probability vector, and*

$$E = \left\{ t \in [0, 1[, \ \lim_{n \to +\infty} \frac{1}{n} \varphi_n(t) = p_k, \forall k \in N_c \right\}.$$

Then,

$$dimE = -\frac{1}{\log c} \sum_{k=0}^n p_k \log p_k.$$

To prove this theorem, we need some preliminary results from probability theory.

Theorem 5.7. *Let $(X_n)_n$ be a sequence of real random variables defined on a probability space $(\Omega, \mathcal{Q}, \mathbb{P})$, such that, the X_n are independent, and with the same law, and $(X_n)_n \in L^2$. Let $M = E(X_1) = \int_\Omega X_1 d\mathbb{P}$. Then*

$$\lim_{n \to +\infty} S_n = \lim_{n \to +\infty} \frac{1}{n} \sum_{i=1}^n E(X_i) \longrightarrow M, \ a.e.$$

Proof. It follows from the equality

$$(E)\big((S_n - M)^2\big) = \frac{1}{n^2} Var\Big(\sum_{i=1}^{n} X_i\Big) = \frac{Var(X_1)}{n}.$$

$\square$

Proof of Theorem 5.6. We take in Billingsley's theorem $\nu = m$ the Lebesgue measure, and so, the problem is to look for μ, such that,

$$E \subset \Big\{ t \in [0, 1[\, ; \ \frac{\log \mu(I_n(t))}{\log |I_n|} \longrightarrow D \Big\},$$

where

$$D = -\frac{1}{\log c} \sum_{k=0}^{n} p_k \log p_k.$$

We next consider $(\mathbb{R}, \mathbf{B}_\mathbb{R})$, and assume that $0 < p_k < 1$. Let μ_n be the measure on $\mathbb{R}$, such that,

$$\mu - n(I_{j_1 \ldots j_n}) = p_{j_1} \ldots p_{j_n}.$$

Such a measure exists. Indeed, take the density measure relatively to the Lebesgue one on I, defined by

$$\mu_n(I_{j_1 \ldots j_n}) = \Big(\frac{p_{j_1} \ldots p_{j_n}}{m(I_{j_1 \ldots j_n})} \Big) m(I_{j_1 \ldots j_n}).$$

We get immediately, $\mu_n([0, 1[) = 1$. Hence, it suffices to prove that there exists a sub-sequence $(\mu_{n_k})_k$ convergent to μ, which is equivalent to

$$\mu_{n_k}(I_{j_1 \ldots j_n}) \to_k \mu(I_{j_1 \ldots j_n}),$$

and thus,

$$\mu(\partial I_{j_1 \ldots j_n}) = 0.$$

Indeed, notice that $(\mu_n(I_{j_1 \ldots j_n}))_n$ is a bounded sequence on $\mathbb{R}$. So, there exists a sub-sequence $(\mu_{n_k})_k$, such that, $(\mu_{n_k}(I_{j_1 \ldots j_n}))_k$ is convergent in $\mathbb{R}$. So, denote $\mu(I_{j_1 \ldots j_n})$ its limit. The weak convergence of measures serves to conclude that μ is also a measure. Moreover,

$$\mu(\{\alpha\}) \leq \underline{\lim}_k t_{n_k}(]\ldots[) \leq 2p_{(s \to \infty)}^{n+s} \longrightarrow 0,$$

where p, n, s are, such that, $I_n(s) \ni \alpha$. Now observe that

$$E \subset \Big\{ t \in [0, 1[, \ \lim_{n \to \infty} \frac{\log \mu(I_n(t))}{\log |I_n(t)|} = D \Big\},$$

and that

$$\mu(I_n(t)) = p_0^{N_0(j)} \dots p_{c-1}^{N_{c-1}(j)} = \prod_{k=0}^{c-1} p_k^{\varphi_n^k(t)},$$

where j is, such that, $I_n(t) = I_{j_1 \dots j_n}$. Consequently,

$$\frac{\log \mu(I_n(t))}{\log |I_n(t)|} = -\frac{1}{n \log c} \sum_{k=0}^{c-1} \varphi_n^k(t) \log p_k = -\frac{1}{\log c} \sum_{k=0}^{c-1} \frac{1}{n} \varphi_n^k(t) \log p_k.$$

Next, consider $\psi_n^k(t) = \delta_{j_n, k}$. It holds easily that

$$\frac{1}{n} \varphi_n^k(t) = \frac{1}{n} \sum_{s=1}^{n} \psi_s^k(t) \xrightarrow[n \to \infty]{} p_k.$$

□

5.5   Exercises for Chapter 5

Exercise 1.
Show that δ defined in section is a distance on $\mathcal{F}$.

Exercise 2.
Consider the set $S = (S_1, S_2)$ where the contractions S_1, and S_2 are defined on $\mathbb{R}$ by

$$S_1(x) = \frac{1}{3}x \quad \text{and} \quad S_2(x) = \frac{1}{3}x + \frac{2}{3}.$$

Let F be the unique non-empty compact S-invariant.

(1) Show that $F = C$, the triadic Cantor set.
(2) Compute the Hausdorff dimension of F using Theorem 5.3.

Exercise 3.
Give an example to show that α-Capacity is not additive.

Exercise 4.
Show that if $\{K_n\}$ is an increasing sequence of sets, that is, $K_n \subset K_{n+1}$, for all $n \in \mathbb{N}$, then

$$Cap_\alpha(\cup_n K_n) = \lim_n Cap_\alpha(K_n).$$

Exercise 5.
Prove that, for all E, F subsets of $\mathbb{R}^d$, we have

$$Cap_\alpha(E \cup F) \leq Cap_\alpha(E) + Cap_\alpha(F).$$

Exercise 6.
Consider $K = \{0\} \cup \{1, \frac{1}{2}, \frac{1}{3}, \frac{1}{4}, ...\}$. Show that

$$dim_{\mathcal{M}}(K) = \frac{1}{2}.$$

Exercise 7.
For $0 < \alpha, \beta < 1$, let $K_{\alpha,\beta}$ be the Cantor set obtained as an intersection of the following nested compact sets. $K^0_{\alpha,\beta} = [0, 1]$. The set $K^1_{\alpha,\beta}$ is obtained by leaving the first interval of length α, and the last interval of length β, and removing the interval in between. To get $K^n_{\alpha,\beta}$, for each interval I in $K^{n-1}_{\alpha,\beta}$, leave the first interval of length $\alpha|I|$, and the last interval of length $\beta|I|$, and remove the sub-interval in between. Compute the Minkowski dimension of $K_{\alpha,\beta}$.

Exercise 8.
Develop a proof of Theorem 5.3.

Exercise 9.
Consider the kernel

$$K(x, y) = \frac{G(x, y)}{G(0, y)} = \frac{|y|^{d-2}}{|x-y|^{d-2}},$$

for $x \neq y$ in $\mathbf{R}^d$, and $K(x, x) = \infty$. Let F be any closed set in $\mathbf{R}^d$; $d \geq 3$, and denote

$$Cop_K(F) = \left(\inf_{\nu(F)=1} \int_F \int_F K(x, y) d\nu(x) d\nu(y) \right)^{-1}.$$

Consider next the spherical shell $F_R = \{x \in \mathbf{R}^d : 1 \leq |x| \leq R\}$. Show that $\lim_{R\infty} Cap_K(F_R) = 2$.

Exercise 10.
For a kernel K, we denote Cap_K the associated Capacity, and Cap_K^∞ the asymptotic Capacity defined for a set A by

$$Cap_K^\infty(A) = \inf_{\{X \text{ finite}\}} Cap_K(A \setminus X).$$

For a cube $C = [a_1, a_{n+1}] \times ... \times [a_d, a_{n+d}]$, $n \geq 1$, denote $d(C)$ the distance of the farthest point in C to 0, and $|C|$ the diameter of C. For $A \in \mathbf{Z}^d$, define

$$H_D^\alpha(A) = \inf \sum_j \left(\frac{|C_j|}{d(C_J)} \right)^\alpha,$$

where the inf is over all coverings of A by cubes. Define next the discrete Hausdorff dimension of A, by

$$dim_D(A) = \inf\{\alpha > 0; \; H_D^\alpha(A) < \infty\}.$$

Consider next, for $\alpha > 0$, the kernel

$$K_\alpha(x, y) = \frac{|y|^\alpha}{1 + |x - y|^\alpha}.$$

1. Show that for all $A \subset \mathbf{Z}^d$, and for all $\alpha > \beta > 0$, we have

$$Cap_{K_\alpha}(A) \geq CH_D^\alpha(A),$$

where $C > 0$ is a positive number depending only on α, and β.

2. Show that if $CapK_\alpha{}^\infty(A) > 0$, then, $H_D^\alpha(A) = \infty$.

3. Show that $dim_D(A) = \inf\{\alpha : CapK_\alpha{}^\infty(A) = 0\}$.

Chapter 6

Packing Measure and Dimension

The packing measure, and dimension constitute the second original essays in fractal analysis, and geometry. This chapter constitutes therefore a second essential part of the book. In which, we focus on the notion of packing measure, which is also in the heart of fractal analysis, and geometry. In fractal analysis, we sometimes say that a set, and/or a measure is fractal when its Hausdorff dimension differs from its packing one. In the present chapter, we will provide in details the original construction of such a measure, its different variants, and the concept of the packing dimension associated to it. Besides, we discuss also the concept of Box dimension, called sometimes the Bouligand–Minkowski dimension, and its link with Hausdorff, and especially packing dimension.

More information, examples, and related topics may be found in [Batakis and Heurteaux (2002); Beardon (1965); Ben Mabrouk and Aouidi (2011); Ben Nasr (1994); Ben Nasr, Bhouri and Heurteaux (2002); Billingsley (1960, 1961, 1965); Boyd (1973); Brown et al (1992); Buck (1970, 1973); Choquet (1961); David and Semmes (1997); Edgar (1998, 2008); Eggleston (1949, 1951); Falconer (1985, 1994, 1990); Frostman (1935); Gervais (2009); Heurteaux (1998); Hutchinson (1981); King (1995); Kigami (2001); Lambert (2012); Mignot (1998); Ngai (1997); Olsen (1995); Peitgen et al (1992a,b); Pesin (1997); Peyriere (1992); Rand (1989); Riedi (1995); Robert (2020); Rogers (1970); Spear (1992); Yeh (2014)].

6.1 Bouligand–Minkowski dimension

For a bounded subset E in $\mathbb{R}^d$, and $\epsilon > 0$, denote $N_\epsilon(E)$ the minimum number of closed Balls of diameter ϵ which cover E, and

$$\Delta(E) = \limsup_{\epsilon \to 0} \frac{\log N_\epsilon(E)}{-\log \epsilon}.$$

Denote also

$$E_\epsilon = \left\{ x \in \mathbb{R}^d, \ d(x, E) \leq \epsilon \right\},$$

where d is the Euclidean distance on $\mathbb{R}^d$. We call ϵ-packing of E, a Countable collection of open Balls $(B_j)_{j \in \mathbb{N}}$, centred in E, disjoint, and satisfying also $|B_j| \leq \epsilon$, $\forall j$, where the symbol $|.|$ stands for the diameter.

Let $M_\epsilon(E)$ be the maximum number of Balls, with diameter ϵ, constituting an ϵ-packing of E. For $n \in \mathbb{N}$, let $W_n(E)$ be the number of dyadic cubes of order n which intersect E.

Proposition 6.1. *For any bounded subset $E \subset \mathbb{R}^d$, we have*

$$\Delta(E) = \limsup_{\epsilon \to 0} \frac{\log M_\epsilon(E)}{-\log \epsilon} = \limsup_{\epsilon \to 0} \frac{\log W_n(E)}{-\log \epsilon} = \limsup_{\epsilon \to 0} (d - \frac{\log m(E_\epsilon)}{\log \epsilon}),$$

where m is the Lebesgue measure.

Proof. Denote Δ_1, Δ_2, and Δ_3 the second, third, and fourth quantities, respectively. Let

$$(B(x_j, \frac{\epsilon}{2}))_j, \ 1 \leq j \leq M_\epsilon(E),$$

be an ϵ-packing of E, such that,

$$\bigcup_j B(x_j, \frac{\epsilon}{2}) \subset E_\epsilon.$$

We immediately obtain

$$M_\epsilon(E) K \epsilon^d \leq m(E_\epsilon),$$

for some constant $K > 0$. Consequently,

$$\Delta_1 \leq \Delta_3.$$

Observe now that $\forall x \in E$, there exists a Ball $B(x_j, \frac{\epsilon}{2})$, such that,

$$d(x, B_j) \leq \frac{\epsilon}{2}.$$

Hence,

$$E \subset \bigcup_j B(x_j, \epsilon),$$

(the closed Balls). This implies that

$$N_\epsilon(E) \le M_\epsilon(E),$$

which yields that

$$\Delta \le \Delta_1.$$

Consider now closed Balls $(B_j)_j$, such that,

$$E \subset \bigcup B_j, \text{ and } E_\epsilon \subset \bigcup_j B(x_j, 2\epsilon),$$

and that,

$$(B(x_j, \frac{\epsilon}{2}))_j, \ 1 \le j \le M_\epsilon(E).$$

We thus have

$$m(E_\epsilon) \le N_\epsilon(E)\epsilon^d K',$$

which yields that

$$\Delta_3 \le \Delta.$$

Consider now the cubes $(C_j)_j$ of order n, such that, $C_j \bigcap E \ne \emptyset$. Let also B_j be a Ball with the same center of C_j, and satisfying

$$|B_j| = |C_j| = \frac{\sqrt{d}}{2^n}.$$

We have immediately,

$$E \subset \bigcup_j B_j.$$

Consequently,

$$N_{\frac{\sqrt{d}}{2^n}}(E) \le W_n(E), \text{ and } \frac{\sqrt{d}}{2^n} < \epsilon \le \frac{\sqrt{d}}{2^{n-1}}.$$

Hence,

$$\Delta \le \Delta_2.$$

Finally, for the same cubes, and Balls of the last step, we get

$$\bigcup_n C_j \subset E_{\frac{\sqrt{d}}{2^n}}.$$

Hence,

$$(\frac{1}{2^n})^d W_n(E) \le m(E_{\frac{\sqrt{d}}{2^n}}).$$

As a result,

$$\Delta_2 \le \Delta_3.$$

$$\square$$

Definition 6.1. $\Delta(E)$ is said to be the Bouligand–Minkowski dimension of E, or also the Box dimension of E.

The following proposition resumes some properties of the Bouligand–Minkowski dimension of sets.

Proposition 6.2.

(1) $0 \leq dimE \leq \Delta(E)$.
(2) $\Delta(\{a\}) = 0$.
(3) $E \subset F \Rightarrow \Delta(E) \leq \Delta(F)$.
(4) $0 \leq \Delta(E) \leq d$.
(5) *Whenever* $m(E) > 0$ *we have* $\Delta(E) = d$.
(6) $\Delta(E) = \Delta(\overline{E})$.
(7) $\Delta(\bigcup_n E_n) \neq \sup_n \Delta(E_n)$.

Proof. (1) Let $s > \Delta(E)$, and $\varepsilon > 0$. Let also $(B(x_i, r_i))_{1 \leq i \leq N_\varepsilon(E)}$ be an ε-covering of E with closed balls. We have

$$\mathcal{H}_\varepsilon^s(E) \leq \sum_i (2r_i)^s \leq (2\varepsilon)^s N_\varepsilon(E).$$

As $s > \Delta(E)$, we deduce that there exists $\varepsilon_0 > 0$, such that, $\forall 0 < \varepsilon < \varepsilon_0$, we have

$$\sup_{0 < \eta < \varepsilon} \frac{\log N_\eta(E)}{-\log \eta} < s.$$

Consequently, $\forall 0 < \eta < \varepsilon$, we get

$$\eta^s N_\eta(E) < 1,$$

which yields that

$$\mathcal{H}^s(E) \leq 2^s < \infty.$$

Hence,

$$dimE \leq s, \quad \forall s > \Delta(E),$$

which means that $dimE \leq \Delta(E)$.
(2) For a singleton $\{a\}$, we notice that $N_\varepsilon(\{a\}) = 1$. So that, $\Delta\{a\} = 0$.
(3) It holds obviously that

$$N_\varepsilon(E) \leq N_\varepsilon(F), \quad \forall \varepsilon > 0.$$

Consequently, $\Delta(E) \leq \Delta(F)$.
(4) Let $R > 0$ be such that $E \subset B(0, R)$, and $\varepsilon > 0$. It holds that

$$\varepsilon^d N_\varepsilon(E) \leq C_d R^d,$$

with some constant $C_d > 0$ depending only on d. This yields that

$$\frac{\log N_\varepsilon(E)}{-\log \varepsilon} \leq d + \frac{\log C_{d,R}}{\log \varepsilon},$$

with some constant $C_{d,R} > 0$ depending only on d and R. As a consequence, we obtain $\Delta(E) \leq d$. The positivity follows from the definition of $\Delta(E)$.
(5) We notice easily, for all $\varepsilon > 0$, that

$$m(E) \leq \sum_{i=1}^{N_\varepsilon(E)} m(B(x_i, r_i)) \leq C_d N_\varepsilon(E) \varepsilon^d.$$

Therefore,

$$d - \frac{\log m(E)}{\log \varepsilon} \leq \frac{\log C_d}{-\log \varepsilon} + \frac{\log N_\varepsilon(E)}{\log \varepsilon},$$

which means that $\Delta(E) \geq d$. By assertion 4, we get the equality.
(6) We obviously have $\Delta(E) \leq \Delta(\overline{E})$. So, we shall prove the opposite inequality. Let $\varepsilon > 0$. We get

$$\overline{E} \subset \bigcup_{i=1}^{N_\varepsilon(E)} B(x_i, 2r_i).$$

Therefore,

$$N_{2\varepsilon}(\overline{E}) \leq N_\varepsilon(E),$$

which yields that $\Delta(E) \geq \Delta(\overline{E})$.
(7) It suffices to take $E = \mathbb{Q} \cap [0, 1]$. $\qquad\square$

6.2 Packing measure

For $E \subset \mathbb{R}^d$, $\alpha > 0$, and $\epsilon > 0$ denote

$$P_\epsilon^\alpha(E) = \sup \sum_j |B_j|^\alpha,$$

where the upper bound is taken over all ϵ-packings $(B_j)_j$ of E. Recall that a Countable family of open Balls $(B_j)_j$ is said to be an ϵ-packing of E if it satisfies $|B_j| \leq \epsilon$, $\forall j$, and $B_j \cap B_k = \emptyset$, $\forall j \neq k$. Denote next

$$P^\alpha(E) = \lim_{\epsilon \to 0} P_\epsilon^\alpha(E).$$

We have the following properties.

Proposition 6.3.

(1) *If E is not bounded, then,*

$$P^\alpha(E) = +\infty, \ \forall \alpha.$$

(2) *If E is bounded, the, $P^d(E) < +\infty$.*
(3) *If $E \subset F$, then,*

$$P^\alpha(E) \le P^\alpha(F), \ \forall \alpha.$$

(4) *For all $\forall E, F \subset \mathbb{R}^d$, we have*

$$P^\alpha(E \bigcup F) \le P^\alpha(E) + P^\alpha(F).$$

(5) *$\forall E, F$, such that, $d(E, F) > 0$, we have*

$$P^\alpha(E \bigcup F) = P^\alpha(E) + P^\alpha(F).$$

(6) *For all subset $E \subset \mathbb{R}^d$, the map $\alpha \mapsto P^\alpha(E)$ satisfies*

$$P^\alpha(E) < \infty \ \Rightarrow \ P^\beta(E) = 0, \forall \beta > \alpha.$$

Proof. (1) Let $\varepsilon > 0$, and consider the cubes $C_{i,n} = [\dfrac{i}{n^{1/\alpha}}, \dfrac{i+1}{n^{1/\alpha}}]$, $i \in \mathbb{N}$. We immediately obtain

$$P_\varepsilon^\alpha(E) \ge \sum_n \frac{1}{n} = \infty.$$

Hence, $P^\alpha(E) = \infty$.
(2) Let $R > 0$ be fixed such that $E \subset B(0, R)$. For any $N_\varepsilon > 0$, and $(B(x_i, r_i))$, $1 \le i \le N_\varepsilon(E)$ and ε-packing of E. We have

$$\sum_i (2\varepsilon)^d \le N_\varepsilon(E).$$

On the other hand,

$$(2\varepsilon)^d N_\varepsilon(E) \le (2R)^d.$$

Consequently,

$$P^d(E) \le (2R)^d < \infty.$$

(3) It follows from the fact that any packing of E is obviously a packing for F.
(4) Let for $\varepsilon > 0$, $(B(x_i, r_i))_i$ be an ε-packing of $E \cup F$, and denote

$$I = \{i \ B(x_i, r_i) \cap F = \emptyset\}, \ \text{and} \ J = \{i \ B(x_i, r_i) \cap E = \emptyset\}.$$

It holds, therefore, that $(B(x_i, r_i))_{i \in I}$ is an ε-packing of E, and $(B(x_i, r_i))_{i \in J}$ is an ε-packing of F. As a result,

$$\sum_i (2r_i)^\alpha = \sum_{i \in I} (2r_i)^\alpha + \sum_{i \in J} (2r_i)^\alpha \leq P_\varepsilon^\alpha(E) + P_\varepsilon^\alpha(F),$$

which yields that

$$P^\alpha(E \cup F) \leq P^\alpha(E) + P^\alpha(E).$$

(5) Let $0 < \varepsilon < \frac{d(E,F)}{2}$. Let also $(B(x_i, r_i))_{i \in I}$ be an ε-packing of E, and $(B(x_i, r_i))_{i \in J}$ an ε-packing of F. It holds that $(B(x_i, r_i))_{i \in I \cup J}$ is an ε-packing of $E \cup F$. Therefore,

$$\sum_{i \in I} \leq (2r_i)^\alpha + \sum_{i \in J} \leq (2r_i)^\alpha \leq P_\varepsilon^\alpha(E \cup F),$$

which yields that

$$P^\alpha(E) + P^\alpha(E) \leq P^\alpha(E \cup F).$$

By assertion 4, we get the equality.

(6) For any ε-packing $(B(x_i, r_i))_i$ of E, and any $\beta > \alpha$, we have

$$\sum_i \leq (2r_i)^\beta \leq (2\varepsilon)^{\beta - \alpha} \sum_i \leq (2r_i)^\alpha.$$

As a consequence,

$$P_\varepsilon^\beta(E) \leq (2\varepsilon)^{\beta - \alpha} P_\varepsilon^\alpha(E).$$

As $P^\alpha(E) < \infty$, we get $P^\beta(E) = 0$. $\qquad\square$

Proposition 6.4. $\forall E \subset \mathbb{R}^d$, there exists a unique constant $D \in \overline{\mathbb{R}}$, such that,

$$P^\alpha(E) = \infty, \ \forall \alpha < D, \ \text{ and } \ P^\alpha(E) = 0, \ \forall \alpha > D.$$

Furthermore, if E is bounded, we have $\Delta(E) = D$.

Proof. Whenever E is not bounded, we put $D = +\infty$. For E bounded, it holds from Proposition 6.3 above, that,

$$\{\alpha \geq 0, \ P^\alpha(E) = 0\} \neq \emptyset.$$

We thus put

$$D = \inf\{\alpha \geq 0, \ P^\alpha(E) = 0\} = \sup\{\alpha \geq 0, \ P^\alpha(E) = \infty\}.$$

Consider next the quantity $M_\epsilon(E)$ defined previously. We obtain

$$M_\epsilon(E)\epsilon^\alpha \leq P_\epsilon^\alpha(E), \ \forall \alpha > 0.$$

In particular, if $\alpha > D$, we get

$$\overline{\lim_{\epsilon \to 0}} M_\epsilon(E)\epsilon^\alpha = 0.$$

Hence,

$$\Delta(E) < \alpha, \ \forall \alpha > D \Rightarrow \Delta(E) \leq D.$$

Let now $0 < \alpha < D$. Then,

$$P^\alpha(E) = +\infty.$$

Consequently,

$$P^\alpha_\epsilon(E) = +\infty, \ \text{for some} \ \epsilon, \ 0 < \epsilon < 1.$$

As a result, there exists, $(B_j)_j$, an ϵ-packing of E, such that,

$$\sum_j |B_j|^\alpha \geq 2.$$

Let next for $n \in \mathbb{N}$, A_n be the number of the B_j's, such that,

$$\frac{1}{2^{n+1}} \leq |B_j| < \frac{1}{2^n}.$$

We have

$$2 \leq \sum_j |B_j|^\alpha \leq \sum_{n=0}^\infty A_n (\frac{1}{2^n})^\alpha.$$

So, whenever $0 < s < \alpha < D$, we get

$$(1 - \frac{1}{2^s}) \sum_{n=0}^\infty (\frac{1}{2^s})^n = 1.$$

Consequently, for

$$A_n (\frac{1}{2^n})^\alpha \geq (\frac{1}{2^s})^n (1 - \frac{1}{2^s}),$$

we immediately observe that

$$(2^{\alpha-s})^n (1 - \frac{1}{2^s}) \leq A_n \leq M_{\frac{1}{2^{n+1}}}(E).$$

Hence,

$$\Delta(E) \geq \alpha - s, \ \forall s > 0,$$

which yields that

$$\Delta(E) \geq \alpha, \ \forall \alpha < D.$$

Consequently,

$$\Delta(E) \geq D.$$

$\square$

Corollary 6.1. P^α *is not an outer measure on* $\mathbb{R}^d$.

Proof. Let $(E_n)_n$ be a collection of subsets of $\mathbb{R}^d$, such that, $\bigcup_n E_n$ is bounded. Let $\alpha > \sup_n \Delta(E_n)$. We get

$$P^\alpha(E_n) = 0, \ \forall n.$$

So, whenever P^α is an outer measure, there holds that

$$P^\alpha(\bigcup_n E_n) = 0.$$

Consequently,

$$\Delta(\bigcup E_n) \geq \alpha.$$

However,

$$\Delta(\bigcup_n E_n) = \sup_n \Delta(E_n) < \alpha,$$

which is contradictory. $\square$

To get an outer measure, we put for $E \subset \mathbb{R}^d$,

$$\hat{P}^\alpha(E) = \inf\left\{\sum P^\alpha(E_n); \ E \subset \bigcup_n E_n\right\},$$

where $(E_n)_n$ is any covering of E with subsets E_n, $n \geq 0$.

Proposition 6.5.

(1) *It holds that,* $\forall \alpha \geq 0$,

$$\hat{P}^\alpha(E) = \inf\left\{\sum_n P^\alpha(E_n) : \ E = \bigcup_n E_n, \ and \ E_n \ is \ bounded \ for \ all \ n\right\}.$$

(2) $\hat{P}^\alpha(E) \leq P^\alpha(E)$, $\forall E \subset \mathbb{R}^d$, *and* $\forall \alpha \geq 0$.
(3) *For all subset* $E \subset \mathbb{R}^d$, *the map* $\alpha \mapsto \hat{P}^\alpha(E)$ *satisfies*

$$\hat{P}^\alpha(E) < +\infty \implies \hat{P}^\beta(E) = 0, \ \forall \beta > \alpha.$$

(4) $\hat{P}^\alpha(E) = 0$, $\forall \alpha > d$, *and* $\forall E \subset \mathbb{R}^d$.
(5) $\hat{P}^\alpha$ *is an outer metric measure on* $\mathbb{R}^d$, $\forall \alpha \geq 0$.

Proof. (1) It is obvious as E is a covering of itself.
(2) Denote $\widetilde{P}^\alpha(E)$ the right hand quantity. We obviously observe that

$$\hat{P}^\alpha(E) \leq \widetilde{P}^\alpha(E).$$

On the other hand, let $(E_n)_n$ be an arbitrary covering of E, and denote, for all n, $F_n = E_n \cap C_n$, where C_n is a cube of side $[-n, n]$ that intersects E_n. We get a covering of E by a sequence of bounded sets $(F_n)_n$. Therefore,

$$\widetilde{P}^\alpha(E) \leq \sum_n P^\alpha(F_n) \leq \sum_n P^\alpha(E_n).$$

Consequently,

$$\widetilde{P}^\alpha(E) \leq \widehat{P}^\alpha(E).$$

(3) As $\widehat{P}^\alpha(E) < \infty$, there exists a covering $(E_i)_i$ of E such that $P^\alpha(E_i) < \infty$, for all i. Consequently, for $\beta > \alpha$, we get from Proposition 6.4, $P^\beta(E_i) = 0$, for all i. As a result, $P^\beta(E_i) = 0$.

(4) Let $\alpha > d$, and $(E_n)_n$ be a covering of E by bounded sets. It follows from Proposition 6.3, that $P^\alpha(E_n) = 0$, for all n. Consequently, due to the fact that

$$\widehat{P}^\alpha(E) \leq \sum_n P^\alpha(F_n),$$

we get $\widehat{P}^\alpha(E) = 0$.

(5) The inequality $\widehat{P}^\alpha(\emptyset) = 0$ is trivial. Let $A \subseteq B$ two subsets in $\mathbb{R}^n$. We get

$$\widehat{P}^\alpha(A) = \inf_{A \subseteq \bigcup_i E_i} \sum_i P^\alpha(E_i) \leq \inf_{B \subseteq \bigcup_i E_i} \sum_i P^\alpha(E_i) = \widehat{P}^\alpha(B).$$

Let now $\varepsilon > 0$, and $(A_n)_n$ be a countable collection of subsets in $\mathbb{R}^d$. There exists $(E_{in})_{i,n}$ such that

$$\widehat{P}^\alpha(A_n) \leq \sum_i P^\alpha(E_{in}) \leq \widehat{P}^\alpha(A_n) + \frac{\varepsilon}{2^n}.$$

It results that

$$\widehat{P}^\alpha(\bigcup_n A_n) \leq \sum_{i,n} P^\alpha(E_{in}) \leq \sum_n \widehat{P}^\alpha(A_n) + \varepsilon.$$

We now prove that $\widehat{P}^\alpha$ is metric. Let $A \subseteq B$. We know that

$$P^\alpha(A) \leq P^\alpha(B).$$

Let $(E_i)_i$ be a covering of $A \cup B$. We have

$$\widehat{P}^\alpha(A) + \widehat{P}^\alpha(B) \leq \sum_i \left(P^\alpha(E_i \cap A) + P^\alpha(E_i \cap A) \right)$$

$$\leq \sum_i P^\alpha(E_i \cap (A \cup B))$$

$$\leq \sum_i P^\alpha(E_i).$$

This yields that

$$\widehat{P}^\alpha(A) + \widehat{P}^\alpha(B) \le \widehat{P}^\alpha(A \cup B).$$

The equality results from the sub-additivity of $\widehat{P}^\alpha$.   □

Definition 6.2. $\widehat{P}^\alpha$ is said to be the packing measure of dimension α on $\mathbb{R}^d$.

6.3   Packing dimension

Let $E \subset \mathbb{R}^d$ be bounded. As for the cases of $\mathcal{H}^\alpha(E)$, and $\mathcal{P}^\alpha(E)$, there exists, for $\widehat{\mathcal{P}}^\alpha(E)$ also, a cutting-off value $D \in \overline{\mathbb{R}}$, satisfying

$$\widehat{\mathcal{P}}^\alpha(E) = +\infty, \text{ for } \alpha < D, \text{ and } \widehat{\mathcal{P}}^\alpha(E) = 0, \text{ for } \alpha > D.$$

Definition 6.3. The critical value D is known as the packing dimension of E, denoted $DimE$. Figure 6.1 illustrates the concept of the packing dimension graphically.

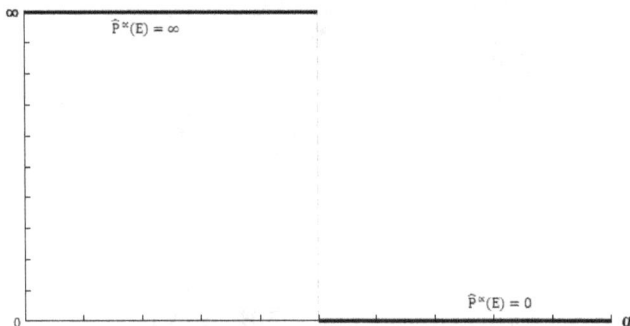

Fig. 6.1: The packing dimension of a set E.

Proposition 6.6.

(1) $0 \le DimE \le d, \forall E \subset \mathbb{R}^d$.

(2) $Dim\{a\} = 0, \forall a \in \mathbb{R}^d$.

(3) $E \subset F \Rightarrow DimE \le DimF, \forall E, F \subset \mathbb{R}^d$.

(4) $Dim\bigcup_n E_n = \sup_n DimE_n, \forall (E_n)_n$ *a sequence of subsets of* $\mathbb{R}^d$.

(5) $\forall E \subset \mathbb{R}^d$, E *is Countable* $\Rightarrow$ $DimE = 0$.

(6) $\forall E \subset \mathbb{R}^d$, $m(E) > 0$, $\Rightarrow DimE = d$, *where* m *stands for the Lebesgue measure on* $\mathbb{R}^d$.

(7) $\forall E \subset \mathbb{R}^d$, E *is bounded* $\Rightarrow$ $DimE \leq \Delta(E)$.

(8) $\forall E \subset \mathbb{R}^d$, $0 < \widehat{P}^\alpha(E) < \infty$ $\Rightarrow$ $\alpha = DimE$.

Proof. (1) $DimE \geq 0$, by definition. From Proposition 6.5, we obtain, for all $\alpha > d$, $\widehat{P}^\alpha(E) = 0$. Consequently,

$$DimE \leq \alpha, \quad \forall \alpha > d,$$

which yields that $DimE \leq d$.

(2) We observe easily that for all $\alpha > 0$, $\widehat{P}^\alpha(\{a\}) = 0$. Consequently, $Dim\{a\} = 0$.

(3) For $E \subset F$, we have $\widehat{P}^\alpha(E) \leq \widehat{P}^\alpha(F)$. Consequently, for $\alpha > DimF$, we get $\widehat{P}^\alpha(E) = 0$.

$$DimE \leq \alpha, \quad \forall \alpha > DimF,$$

which yields that $DimE \leq DimF$.

(4) From the previous assertion, we get

$$\sup_n DimE_n \leq Dim \bigcup_n E_n.$$

Conversely, for all $\alpha > \sup_n DimE_n$, we have

$$\widehat{P}^\alpha(\bigcup_n E_n) \leq {}_n\widehat{P}^\alpha(E_n) = 0.$$

Consequently,

$$\alpha \geq Dim \bigcup_n E_n, \quad \forall \alpha > \sup_n DimE_n.$$

As a result,

$$Dim \bigcup_n E_n \leq \sup_n DimE_n.$$

(5) Write $E = \{a_n, \ n \in \mathbb{N}\}$. We get from previously,

$$DimE = \sup_n Dim\{a_n\} = 0.$$

(6) Let $(E_n)_n$ be a covering of E. We have,

$$0 < m(E) \leq \sum_n m(E_n).$$

Therefore, there exists $n_0 \in \mathbb{N}$, such that, $m(E_{n_0}) > 0$. Consequently, for all ε-packing $(B(x_i, r_i))_i$ of E_{n_0}, we have

$$0 < m(E_{n_0}) \leq \sum_i m(B(x_i, r_i)) \leq C \sum_i (2r_i)^d \leq C(2\varepsilon)^{d-\alpha} \sum_i (2r_i)^\alpha.$$

As a consequence, $P^\alpha(E_{n_0}) = \infty$. As a result,

$$\sum_i P^\alpha(E_n) = \infty,$$

for all coverings $(E_n)_n$ of E. This yields that $\widehat{P}^\alpha(E) = \infty$, for all $\alpha < d$. Consequently, $DimE \geq d$. By assertion 1, we get the equality.

(7) For $\alpha > \Delta(E)$, we get $P^\alpha(E) = 0$, which yields that $\widehat{P}^\alpha(E) = 0$. As a result, $DimE \leq \Delta(E)$.

(8) $P^\alpha(E) > 0$ yields that $\alpha \leq DimE$. Similarly, $P^\alpha(E) < \infty$ yields that $\alpha \geq DimE$. Consequently, $\alpha = DimE$.   □

Proposition 6.7. $\forall E \subset \mathbb{R}^d$, we have

$$DimE = \inf\left\{\sup_n \Delta(E_n), \ E = \bigcup_n E_n, \ E_n \ bounded, \ \forall n\right\}.$$

Proof. Let $\widehat{\Delta}(E)$ be the right hand term. We have

$$\Delta(E_n) \geq DimE_n, \ \forall n.$$

As a consequence,

$$\sup_n \Delta(E_n) \geq \sup_n DimE_n = DimE.$$

Hence,

$$\widehat{\Delta}(E) \geq DimE.$$

Let next, $\alpha > DimE$, then,

$$\widehat{P}^\alpha(E) = 0.$$

Therefore, there exists bounded subsets E_n, $n \in \mathbb{N}$, such that,

$$E = \bigcup_n E_n,$$

and

$$\sum_n P^\alpha(E_n) < \infty.$$

As a result,

$$\Delta(E_n) \leq \alpha, \ \ \forall n.$$

Thus,

$$\hat{\Delta}(E) \leq \alpha, \quad \forall \alpha > DimE,$$

which yields that

$$\hat{\Delta}(E) \leq DimE.$$

$\square$

Corollary 6.2. $\forall E \subset \mathbb{R}^d$, we have

$$dimE \leq DimE.$$

Proof. Write $E = \bigcup_n E_n$. We have,

$$dimE_n \leq \Delta(E_n).$$

Hence, $\forall n$, we get

$$dimE = \sup_n dimE_n \leq \sup_n \Delta(E_n) = DimE.$$

$\square$

6.4   Exercises for Chapter 6

Exercise 1.
Let $E_1 \subset \mathbb{R}^{d_1}, E_2 \subset \mathbb{R}^{d_2}$. Show that

$$dimE_1 \times E_2 \leq dimE_1 + DimE_2,$$

and that

$$DimE_1 \times E_2 \leq DimE_1 + DimE_2.$$

Exercise 2.
Let $E_1 \subset \mathbb{R}^{d_1}$ and $E_2 \subset \mathbb{R}^{d_2}$, where $d_1, d_2 \in \mathbb{N}$. Show that, if E_2 is bounded, we have

$$dimE_1 \times E_2 \leq dimE_1 + \Delta(E_2),$$

and

$$\Delta(E_1 \times E_2) \leq \Delta(E_1) + \Delta(E_2).$$

Exercise 3.
Let C be the triadic Cantor set. Show that

$$DimC = \frac{\log 2}{\log 3}.$$

Exercise 4.

For $k \in \mathbb{N}$, let $\{X_{i_1,\ldots,i_k} : i_j = 1 \quad or \quad 2\}$, be the set of 2^k intervals of length 3^{-k} that occur in the usual construction of the middle-third Cantor set E, at the level k, and nested in the usual way.

(1) Show that, setting

$$\mu(X_{i_1,\ldots,i_k}) = 2^{-k},$$

leads to a measure on E.

(2) Show that the same is true by setting

$$\mu(X_{i_1,\ldots,i_k}) = (1/3)^{n_1}(2/3)^{n_2},$$

where n_1, and n_2 are the number of occurrences of the digits 1, and 2 respectively in $(i_1,\ldots,i_k)$.

Exercise 5.

Let $\{F_1,\ldots,F_m\}$ be an iterated function system consisting of similarity transformations satisfying the open set condition, with attractor E of dimension $s \in \mathbb{R}$. Show that

$$\mathcal{H}^s(F_i(E) \cap F_j(E)) = 0, \quad \text{if} \ \ i \neq j.$$

Exercise 6.

Find the packing, and the box dimensions of the set

$$X = \{(1/p, 1/q) : p, q \in \mathbb{Z}^+\} \subset \mathbb{R}^2.$$

Exercise 7.

Let E be the middle-third Cantor set. Find estimates for $\mathcal{H}^s(E)$, and $\mathcal{P}^s(E)$, where $s = \log 2/\log 3$.

Exercise 8.

Let $0 < \lambda < 1/2$, and $F_1, F_2 : \mathbb{R} \to \mathbb{R}$ be given by

$$F_1(x) = \lambda x, \quad \text{and} \quad F_2(x) = \frac{1}{2}x + \frac{1}{2}.$$

Describe the attractor of $\{F_1, F_2\}$, and find an expression for its Hausdorff, and packing dimensions.

Exercise 9.

(1) Show that the hypotheses of Theorem 3.2 imply that

$$\mathcal{P}^s(E) \leq \mathcal{P}_0^s(E) < \infty,$$

where $\mathcal{P}^s$, and $\mathcal{P}_0^s$ are the packing measure, and pre-measure, (See Section 2.1).

(2) Assume that

$$\mathcal{P}_r^s(E) > a^{-s},$$

and show that, there exists disjoint Balls $B_1, \ldots, B_m$, with

$$\sum |B_i|^t > a^{-t},$$

for $t > s$.

(3) Let $g_i : E \to E \cap B_i$, satisfying

$$a|B_i||x - y| \le |g_i(x) - g_i(y)|.$$

Show that $dim_H E > s$.

Exercise 10.

Let E be the subset of the middle-third Cantor set consisting of those numbers in $[0, 1]$, with base 3 expansion containing only the digits 0, and 2, and where two consecutive digits 2 are not allowed. Let

$$E_1 = E \cap [0, 1/3], \quad \text{and} \quad E_2 = E \cap [2/3, 0].$$

Show that

$$dim_H E_1 = dim_H E_2 = \log((1 + \sqrt{5})/2)/\log 3.$$

Chapter 7

Multifractal Analysis of Gibbs Type Measures

In this chapter, we investigate one of the most important concepts in fractal analysis; the so-called multifractal formalism for measures. This is a main common point with the physical meaning, and interactions with fractals, especially with the concept of Gibbs measures. Indeed, the word multifractal has been firstly introduced in [Frisch and Parisi (1985)], in the framework of statistical physics, by Frisch, and Parisi, when studying turbulence. To reach such a formalism, the multifractal analysis focuses on the singularities of the irregular object, such as the measure, and computes the so-called spectrum of singularities by means of the Hausdorff dimension of some sets associated to the measure, and known as the singularities sets. A simple way to describe the geometry, or the structure of the support of a measure μ is to compute its box dimension, i.e, the number d, such that, $N_\delta(E)$ has the same order as δ^{-d}, where $N_\delta(E)$ is the optimal number of balls with diameter δ necessary to cover the support. For a measure μ, the multifractal formalism is reduced to the statistical characterization of the local scaling invariance of the measure. One covers the support of the measure μ by subsets

$$E(\alpha) = \{x \in \mathbb{R}^m; \ \mu(B(x,r)) \sim r^\alpha\}.$$

The spectrum of singularities will be $f\alpha) = dim\, E(\alpha)$.

Based on some analogy with the statistical physics, the spectrum is evaluated by means of an auxiliary function $\tau(q)$ which has the same rule as the thermodynamical potential, such as the free energy. Let

$$\tau(q) = \frac{\log\left(\sup \sum_i \mu(U_i)^q\right)}{-\log r}.$$

The multifractal formalism affirms that

$$f(\alpha) = \inf_q \left(\alpha q + \tau(q)\right).$$

The sup above is taken over all the countable partitions of $Support(\mu)$ composed with coverings U_i's with diameters at most r. q is the analogous of the temperature inverse in thermodynamics.

In some cases of statistical, and/or thermodynamical physics, the transfer of energy is governed by cascade models, in the sense that, a disturbance on a scale receives energy from a larger one, and transfers it to smaller scale disturbances. This defines a Gibbs measure. Under the assumption that the energy transfer rate is constant both in space, and in the cascade steps, we obtain the scaling law for the velocity

$$|v(t + h) - v(t)|^q \sim |h|^{-q/3}.$$

However, experimentally, there are some intermittency due to possible presence of strong fluctuations of the energy transfer, and dissipation. This breaks down the low cited above for large q. So one shall take into account the non-cascading or the non-self-similar transfer.

In the present chapter, we focus on the study of the multifractal analysis, and the multifractal formalism for the class of Gibbs measures, and show that in this case, the validity of the multifractal formalism is proved naturally. We propose in this section to recall some facts about Gibbs measures, and to explain the idea of computing the spectrum of singularity in some situation, where the construction of such measures is possible. This is mainly based on the possibility of constructing Gibbs measures supported by the singularity set. The main results exposed here are based on [Brown et al (1992)]. However, more formulations, and cases may be found in the whole list [Barnsley (2000); Batakis and Heurteaux (2002); Beardon (1965); Bélair (1987); Ben Mabrouk (2005, 2007, 2008a,b,c,e); Ben Mabrouk and Ben Abdallah (2006); Ben Mabrouk and Aouidi (2011, 2012); Ben Mabrouk, Aouidi and Ben Slimane (2014, 2016); Ben Nasr (1994, 1997); Ben Nasr and Bhouri (1997); Ben Nasr, Bhouri and Heurteaux (2002); Billingsley (1960, 1961, 1965, 1968); Buck (1970, 1973); Choquet (1961); Collet et al (1987); David and Semmes (1997); Edgar (1998, 2008); Eggleston (1949); Falconer (1994, 1985, 1990); Frostman (1935); Frame and Cohen (2015); Frisch and Parisi (1985); Gervais (2009); Gurevich and Tempelman (1999); Heurteaux (1998); Hutchinson (1981); King (1995); Lambert (2012); Lucas (2000); Mandelbrot (1982, 1995, 1997, 1999, 2004); Menceur, Ben Mabrouk and Betina (2016); Menceur and Ben Mabrouk (2019); Mignot (1998); Ngai (1997); Nguyen (2001); Olivier (1998); Olsen (1995); Pesin (1997); Pesin and Climenhaga (2009); Peyriere (1992); Rand (1989); Riedi (1995); Robert (2020); Rogers (1970); Sierpinski (1916); Spear (1992); Triebel (1992); Veerman (1998); Wu (1998); Yeh (2014)].

In the sequel, and for the sake of simplicity, we will restrict to the Euclidean space $\mathbb{R}$. The general case may be deduced in a natural way. Consider the unit interval $[0, 1[$, a sequence of partitions $\mathfrak{F}_n$, $n \in \mathbb{N}$, and denote $\mathfrak{F} = \bigcup_n \mathfrak{F}_n$. We also assume, for all $n \in \mathbb{N}$, that $\mathfrak{F}_{n+1}$ is a refinement of $\mathfrak{F}_n$, and that $\forall t \in [0, 1[$, we have $|I_n(t)| \longrightarrow 0$ as $n \to +\infty$, where $I_n(t)$ is the unique element of $\mathfrak{F}$ that contains t.

We consider the analogues of the Hausdorff measure $\mathcal{H}^\alpha$, the packing pre-measure $\mathcal{P}^\alpha$, and the packing measure $\widehat{\mathcal{B}}^\alpha$, introduced previously, by restricting the coverings, and the packings on elements of the collection $\mathfrak{F}$. We thus obtain, in a similar way, associated dimensions, which will be denoted here also dim, Dim, and Δ. Finally, for a real-valued function φ, we denote φ_L its Legendre transform defined, for $\alpha \in \mathbb{R}$, by

$$\varphi_L(\alpha) = \inf_{x \in \mathbb{R}} (\alpha(x + 1) - \varphi(x)).$$

7.1   The multifractal formalism

The idea to be exposed here is based on [Brown et al (1992)], where the authors constructed a type of partition function, easy to compute, and thus permits to evaluate the spectrum of singularities for measures. For $n \in \mathbb{N}$, $x, y \in \mathbb{R}$, and μ a Borel probability measure on $[0, 1[$, denote

$$C_n(x, y) = \frac{1}{n} \log \sum_{I \in \mathfrak{F}_n} \mu(I)^{x+1} |I|^{-y},$$

where $\sum$ is taken on the elements $I \in \mathfrak{F}_n$, such that, $\mu(I) \neq 0$. Denote next

$$C(x, y) = \limsup_n C_n(x, y) \in \overline{\mathbb{R}}.$$

It is straightforward that C is convex, increasing as a function of y, and decreasing relatively to the variable x. Denote next

$$\Omega = \Big\{ (x, y); \ C(x, y) < 0 \Big\}.$$

Lemma 7.1. *The following assertions hold. (1) There exists a function $\varphi : \mathbb{R} \to \overline{R}$, concave, nondecreasing, and satisfying,*

$$\mathring{\Omega} = \Big\{ (x, y) \in \mathbb{R}^2; \ y < \varphi(x - 0) \Big\},$$

where $\mathring{\Omega}$ stands for the topological interior of Ω.
(2) If φ is finite on an open interval containing -1, then, it is also finite on $[a, +\infty[$, for all $a \in [-1, \infty[$.

Proof. (1) Denote, for $x \in \mathbb{R}$,

$$\varphi(x) = \sup\{y \in \mathbb{R}; \; C(x, y) < 0\}.$$

Let also, $x_1 < x_2$ be two real numbers. We immediately observe that

$$C(x_2, y) < C(x_1, y); \; \forall y \in \mathbb{R}.$$

As a consequence,

$$C(x_2, y) < 0, \; \forall y < \varphi(x_1).$$

Hence,

$$y < \varphi(x_2), \; \forall y < \varphi(x_1).$$

By taking the sup on y, we get

$$\varphi(x_1) \le \varphi(x_2).$$

Let now (x_1, y_1), and (x_2, y_2) be, such that,

$$y_1 < \varphi(x_1) \text{ and } y_2 < \varphi(x_2).$$

For all $t \in [0, 1]$, we have, from the convexity of C,

$$C(tx_1 + (1 - t)x_2, ty_1 + (1 - t)y_2) \le tC(x_1, y_1) + (1 - t)C(x_2, y_2) < 0.$$

This yields that

$$ty_1 + (1 - t)y_2 < \varphi(tx_1 + (1 - t)x_2).$$

Taking the sup on y_1, and y_2, we obtain

$$t\varphi(x_1) + (1 - t)\varphi(x_2) \le \varphi(tx_1 + (1 - t)x_2).$$

Now, denote

$$\widetilde{\Omega} = \left\{(x, y) \in \mathbb{R}^2; \; y < \varphi(x - 0)\right\}.$$

We shall prove that $\overset{\circ}{\Omega} = \widetilde{\Omega}$. It is easy to see that $\widetilde{\Omega}$ is open. If there exists an auxiliary open set O, such that,

$$\widetilde{\Omega} \subsetneqq O \subsetneqq \Omega,$$

we immediately deduce that, there exists a point (x, y), such that,

$$(x, y) \in O, \text{ and } (x, y) \notin \widetilde{\Omega}.$$

This point satisfies easily

$$C(x, y) < 0 \; (\Longleftrightarrow y < \varphi(x)), \text{ and } y \ge \varphi(x) \; (\Longleftrightarrow C(x, y) \ge 0).$$

This is a contradiction.

(2) This follows immediately from the concavity of φ. $\qquad\square$

Denote now

$$B_\alpha = \left\{ t \in [0,1[; \ \limsup_{n \to +\infty} \frac{\log \mu(I_n(t))}{\log |I_n(t)|} \leq \alpha \right\},$$

and

$$V_\alpha = \left\{ t \in [0,1[; \ \limsup_{n \to +\infty} \frac{\log \mu(I_n(t))}{\log |I_n(t)|} \geq \alpha \right\}.$$

The following result is the analogue of the one proved in [Brown et al (1992)].

Theorem 7.1.

(1) *For all $\alpha \leq \varphi'(-1-0)$, we have $DimB_\alpha \leq \varphi_L(\alpha)$.*
(2) *For all $\alpha \geq \varphi'(-1+0)$, we have $DimV_\alpha \leq \varphi_L(\alpha)$, where φ_L is the Legendre transform of φ, defined above.*

Proof. For $p \in \mathbb{N}$, and $\alpha, \eta > 0$, denote

$$B_\alpha(\eta, p) = \left\{ t \in B_\alpha; \ \forall n, \ |I_n(t)| \geq \frac{1}{p} \ \text{ or } \ \mu(I_n(t)) \geq |I_n(t)|^\eta \right\}.$$

Let $t \in B_\alpha$. There exists $q \in \mathbb{N}$, such that, $\forall n \geq q$,

$$\mu(I_n(t)) \geq |I_n(t)|^\eta.$$

Let p be, such that,

$$\frac{1}{p} < \inf \left\{ |I|, \ I \in \bigcup_{j=1}^{q} \mathfrak{F}_j \right\}.$$

It holds, for $t \in B_\alpha(\eta, p)$, with $|I_n(t)| < \frac{1}{p}$, that $n > p$. Thus,

$$\mu(I_n(t)) \geq |I_n(t)|^\eta.$$

Consequently,

$$B_\alpha = \bigcup_n B_\alpha(\eta, p).$$

It suffices then to prove the inequality for $B_\alpha(\eta, p)$. To do it, observe that, for $\{I_j\}$ an ϵ-packing of $B_\alpha(\eta, p)$, and $0 < \epsilon < \frac{1}{p}$, we have

$$\sum_j |I_j|^\delta = \sum_j |I_j|^\delta |I_j|^{\eta t} |I_j|^{-\eta t}, \quad t > 0,$$

$$\leq \sum_j |I_j|^{\delta - \eta t} \mu(I_j)^t$$

$$\leq \sum_j |I_j|^{t+1-1} |I_j|.$$

As a result, the last right hand member is bounded if

$$\eta t - \delta < \varphi(t - 1),$$

which is equivalent to

$$\delta > \eta t - \varphi(t - 1) > 0,$$

which is always possible because $B_\alpha \neq \emptyset$. Hence,

$$\Delta B_\alpha(\eta, p) \leq \eta t - \varphi(t - 1), \; \forall t > 0, \; \forall n > \alpha.$$

Consequently,

$$Dim B_\alpha(\eta, p) \geq \eta t - \varphi(t - 1).$$

Therefore,

$$Dim B_\alpha \leq \eta t - \varphi(t - 1), \; \forall t > 0, \; \forall n > \alpha.$$

Therefore, $\forall \eta > \alpha$, we get

$$\begin{aligned}
Dim B_\alpha &\leq \inf_{t > -1}(\eta t - \varphi(t - 1)) \\
&= \inf(\eta(t + 1) - \varphi(t) \\
&= \inf_{t \geq -1}(\eta(t + 1) - \varphi(t)).
\end{aligned}$$

So, finally,

$$Dim B_\alpha \leq \inf_{t \geq -1}(\alpha(t - 1) - \varphi(t)) = \varphi^*(\alpha).$$

$\square$

Denote for the next

$$E_\alpha = \left\{ t \in [0, 1[, \lim_n \frac{\log \mu(I_n(t))}{\log |I_n(t)|} = \alpha \right\}.$$

The following result is due to [Brown et al (1992)].

Theorem 7.2. *Assume that φ is differentiable at x, and that there exists a Borel probability measure μ_x on $[0, 1[$, such that, $\forall \, I \in \mathfrak{F}$,*

$$\frac{1}{M} \mu(I)^{x+1} |I|^{-\varphi(x)} \leq \mu_x(I) \leq M \mu(I)^{x+1} |I|^{-\varphi(x)}.$$

Then,

$$dim E_{\varphi'(x)} = \varphi^*(\varphi'(x)).$$

The measure μ_x is called Gibbs measure.

Proof. We show firstly that μ_x is supported on $E_{\varphi'(x)}$. Next, we apply Billingsley Theorem to μ_x. Indeed, denote

$$V_\alpha^* = \left\{ t \in [0,1[, \liminf_{n \to \infty} \frac{\log \mu(I_n(t))}{\log |I_n(t)|} \geq \alpha \right\}.$$

We will show that

$$\mu_x(V_{\varphi'(x)}^*) = \mu_x(B_{\varphi'(x)}) = 1.$$

Observe that

$$V_\alpha^* \bigcap B_\alpha = E_\alpha.$$

So, for $\alpha < \varphi'(x)$, we get $\mu_x(V_\alpha^*) = 1$. Assume now that $\varphi'(x) > 0$, and let α be, such that,

$$0 < \alpha < \frac{\varphi(x+t) - \varphi(x)}{t} < \varphi'(x).$$

For $n \in \mathbb{N}$, consider also,

$$A_n = \{ t \in [0,1[, \mu(I_n(t)) > |I_n(t)|^\alpha \}.$$

We have

$$\mu_x(A_n) = \sum_{I \in \mathfrak{F}_n; \mu(I) > |I|^\alpha} \mu_x(I) |I|^{\alpha t} |I|^{-\alpha t}, \quad (t > 0)$$

$$\leq K \sum_{I \in \mathfrak{F}_n; \mu(I) > |I|^\alpha} \mu(I)^{x+1} |I|^{-\varphi(x)} \mu(I)^t |I|^{-\alpha t}$$

$$\leq K \sum_{I \in \mathfrak{F}_n} \mu(I)^{x+t+1} |I|^{-(\varphi(x)+\alpha t)}.$$

Let next t be, such that,

$$\varphi(x) + \alpha t < \varphi(x+t),$$

or equivalently,

$$\alpha < \frac{\varphi(x+t) - \varphi(x)}{t},$$

which is possible, because

$$\alpha < \varphi'(x) = \lim_{t \to 0} \frac{\varphi(x+t) - \varphi(x)}{t}. \qquad \square$$

Theorem 7.3. *Let x be, such that, $\varphi'(x)$ exists. The following assertions hold.*

(1) $\mu_x(I) \leq M\mu(I)^{x+1} |I|^{-\varphi(x)}$ *if* $\mu(x) \neq 0$.
(2) $\mu_x(I) = 0$ *for* $\mu(x) = 0$.
(3) $dim E_{\varphi'(x)} = \varphi^*(\varphi'(x)) = (x+1)\varphi'(x) - \varphi(x)$.

Remark 7.1.

(1) Theorem 7.3 is valid if φ is finite on the interval J without -1.
(2) If φ is differentiable at x, and If $(x+1)\varphi'(x) - \varphi(x) = 0$, then, $dim E_{\varphi'(x)} = 0$.

7.2 Existence of Gibbs measures

Let $\mathbb{A}$ be the set of words composed from $\{0, 1, 2, \dots, p-1\}$ as an alphabet, $p \geq 2$ in $\mathbb{N}$. For $n \in \mathbb{N}$, let $\mathbb{A}_n$ be the set of elements of $\mathbb{A}$ whom length equals n. Let next μ be a measure on $\mathcal{A}$, and $\mathfrak{F} = \{I_a \in \mathcal{A}_n\}_{n \geq 1}$ a partition of $[0, 1]$. Denote $\ell(a) = |I_a|$. Assume finally that μ, and ℓ satisfy respectively

$$\frac{1}{M}\mu(a)\mu(b) \leq \mu(ab) \leq M\mu(a)\mu(b),$$

and

$$\frac{1}{L}l(a)l(b) \leq l(ab) \leq l(a)l(b),$$

for some constants $L, M > 0$ fixed. We have the following result.

Proposition 7.1. *Denote for* $x, y \in \mathbb{R}$,

$$C_n(x, y) = \frac{1}{n} \log \sum_{a \in \mathbb{A}_n}^{*} \mu(a)^{x+1} l(a)^{-y},$$

and $C(x, y) = \lim_n C_n(x, y)$. *Then,* $C(x, y)$ *is finite,* $\forall x, y \in \mathbb{R}$.

Proof. Denote, for $n \in \mathbb{N}$,

$$Z_n = \sum_{a \in \mathbb{A}_n} \mu(a)l(a).$$

Hence,

$$C_n(0, -1) = \frac{1}{n} \log Z_n, \ \forall n \in \mathbb{N}.$$

Next, observe that

$$Z_{n+m} = \sum_{ab, a \in \mathbb{A}_m, b \in \mathbb{A}_n} \mu(ab)l(ab), \ \forall n, m \in \mathbb{N}.$$

We immediately obtain

$$\frac{1}{ML} Z_n Z_m \leq Z_{n+m} \leq ML Z_n Z_m, \ \forall n, m \in \mathbb{N}.$$

This may be written as

$$|\log Z_{n+m} - \log Z_n - \log Z_m| \leq \log ML, \ \forall n, m \in \mathbb{N}.$$

Denoting, for $n \in \mathbb{N}$, $U_n = \log Z_n$, we get

$$|U_{m+n} - U_n - U_m| \leq K, \ \forall n, m \in \mathbb{N}.$$

This yields that for some $\ell \in \mathbb{R}$, we have

$$|\frac{U_n}{n} - l| \leq \frac{K}{n}, \ \forall n \in \mathbb{N}.$$

Consequently,

$$|C_n - C| \leq \frac{\log ML}{n}, \ \forall n \in \mathbb{N}.$$

$\square$

Denote for the next,

$$l'(a) = \mu(a)^x l(a)^{-y}.$$

It follows from Proposition 7.1 that, for $\mu(a) \neq 0$, we obtain

$$\frac{1}{K_1} l'(a) l'(b) \leq l'(ab) \leq K_1 l'(a) l'(b).$$

We have the following result.

Proposition 7.2. *Whenever*

$$\lim_{n \to +\infty} \left(\sup_{a \in \mathbb{A}} \mu(a) l(a) e^{-nc} \right) = 0,$$

there exists a Borel probability measure γ on $[0, 1[$, and a constant $K \geq 1$, such that,

$$\frac{1}{K} \mu(a) l(a) e^{-nc} \leq \gamma(a) \leq K \mu(a) l(a) e^{-nc}, \quad \forall a \in \mathbb{A}_n.$$

Denote for $n \in \mathbb{N}$,

$$\mathbb{A}_n^* = \{a \in \mathbb{A}_n, \mu(a) \neq 0\}.$$

It is obvious that in the proposition, we may restrict to A_n^* instead of A_n. Whenever $l(a)$ is not defined for $a \in \mathbb{A}_n$, such that, $\mu(a) = 0$, the hypothesis $a \in \mathbb{A}_n^*$ is necessary. In the other cases, $\gamma(a) = 0$, whenever $\mu(a) = 0$, the proposition may be expressed in another way.

Proposition 7.3. *Assume that*

$$\lim_{n \to \infty} \left[\sup_{a \in \mathbb{A}_n^*} \mu(a) l(a) e^{-nc} \right] = 0, \quad \forall a \in \mathbb{A}.$$

There exists a probability measure γ on $[0, 1[$, and a constant $K \geq 1$, such that,

(1) $\frac{1}{K} \mu(a) l(a) e^{-nc} \leq \gamma(a) \leq K \mu(a) l(a) e^{-nc}, \forall a \in \mathbb{A}_n^*.$
(2) $\forall a \in \mathbb{A}, \mu(a) = 0 \implies \gamma(a) = 0.$

Proof. Let, for $n \in \mathbb{N}$, l_n be the function defined on $[0, 1[$ by $l_n(t) = |I_n(t)|$, where for $t \in [0, 1[$, $I_n(t)$ is the interval of $\mathcal{A}_n$ containing t. For $s \in \mathbb{R}$, denote

$$\varphi_s = \sum_{n=1}^{\infty} l_n e^{-ns}.$$

We observe that

$$\int l_n d\mu = \sum_{a \in \mathbb{A}_n} \mu(a) l(a) = Z_n,$$

and

$$\int \varphi_s d\mu = \sum_{n=1}^{\infty} Z_n e^{-ns}.$$

Denote next for $s > c$,

$$Z(s) = \int \varphi_s d\mu < \infty.$$

We obtain a Borel probability measure P_s on $[0, 1[$, expressed, for any Borel set $B \subset [0, 1[$, as

$$P_s(B) = \frac{1}{Z(s)} \int_B \varphi_s d\mu.$$

Let next $a = t_1...t_n \in \mathbb{A}_n$ be fixed, and denote for $1 \le j \le n$, $a_j = t_1...t_j$. We have, for $\mu(a) \neq 0$,

$$Z(s) P_s(a) = \int_{I_a} \varphi_s d\mu$$

$$= \sum_{j=1}^{\infty} (\int_{I_a} l_j d\mu) e^{-js}$$

$$= \sum_{j=1}^{n} (\int_{I_a} l_j d\mu) e^{-js} + \sum_{j=n+1}^{\infty} (\int_{I_a} l_j d\mu) e^{-js}.$$

Whenever $j \ge n + 1$, we get

$$\int_{I_a} l_j d\mu = \sum_{b \in \mathbb{A}_{j-n}} (\int_{I_{ab}} l_j d\mu) = \sum_{b \in \mathbb{A}_{j-n}} l(ab) \mu(ab).$$

For $1 \le j \le n$, we have similarly

$$\int_{I_a} l_j d\mu = l(ab) \mu(a).$$

As a result,

$$Z(s) P_s(a) = \mu(a) \sum_{j=1}^{n} l(a_j) e^{-js} + \sum_{j=1}^{\infty} \sum_{b \in \mathbb{A}_j} l(ab) \mu(ab) e^{-(n+j)s}.$$

Now, observe that

$$\sum_{b \in \mathbb{A}_j} l(ab) \mu(ab) = \left(\sum_{b \in \mathbb{A}_j} \frac{l(ab)}{l(a)} \frac{\mu(ab)}{\mu(a)} \right) l(a) \mu(a).$$

Denote

$$Z_{(n,a)} = \sum_{b \in \mathbb{A}_j} \frac{l(ab)}{l(a)} \frac{\mu(ab)}{\mu(a)},$$

and

$$Z_a(s) = \sum_{j=1}^{\infty} Z_j e^{-js}.$$

We obtain

$$Z(s)P_s(a) = \mu(a) \sum_{j=1}^{n} l(a_j) e^{-js} + l(a)\mu(a) e^{-ns} Z_a(s).$$

On the other hand,

$$\frac{K^{-2} e^{c-s}}{1 - e^{c-s}} \le Z_a(s) \le \frac{K^2 e^{c-s}}{1 - e^{c-s}}, \ \forall a,$$

and

$$\frac{K^{-2} e^{c-s}}{1 - e^{c-s}} \le Z(s) \le \frac{K^2 e^{c-s}}{1 - e^{c-s}}, \ \forall a.$$

Consequently,

$$\frac{1}{n} \log Z_{(n,a)} \longrightarrow C \quad as \quad n \longrightarrow +\infty.$$

We now emphasize the case where $\mu(a) = 0$, and $P_s(a) = 0$. Indeed, P_s is considered as a probability measure on $\mathbb{R}$. We have $P_s([0,1]) = 1$. Let $(s_k)_k \subset \{s > c\}$ be, such that, $s_k \xrightarrow[k \to +\infty]{} c$. Hence, (P_{s_k}) is weakly convergent to a probability measure γ on $\mathbb{R}$. Consequently, $\forall a$,

$$P_{s_k}(a) \xrightarrow[k \to +\infty]{} \gamma(a).$$

It remains to prove that $\gamma(\partial I_a) = 0$. To do it, write $a = t_1...t_p$. Let $t \in \mathring{I}_a$, and I_b, I_c be, such that, $z \in I_b \cup I_c \subset \mathring{I}_a$. Denote finally $\tilde{I} = I_b \mathring{\cup} I_c$. We have

$$\gamma(z) \le \gamma(]\tilde{I}[) \le \lim_{k \to +\infty} P_{s_k}(]\tilde{I}[).$$

On the other hand,

$$\begin{aligned}
p_{s_k}(]\tilde{I}[) &\le p_{s_k}([I_b \cup I_c[) \\
&= p_{s_k}(I_b) + p_{s_k}(I_c) \\
&\le \frac{\mu(b)}{Z(b)} \sum_{j=1}^{n} l(b_j) e^{-js_k} + K_1 \mu(b) l(b) e^{-ns_k} \\
&\quad + \frac{\mu(c)}{Z(c)} \sum_{j=1}^{n} l(c_j) e^{-js_k} + K_1 \mu(c) l(c) e^{-ns_k}.
\end{aligned}$$

Consequently,

$$\gamma(z) \le \gamma(]\tilde{I}[) \le K_1 \Big(\mu(b) l(b) e^{-nc} + \mu(c) l(c) e^{-nc} \Big).$$

$\square$

Remark 7.2. Denote, for $n \geq 1$,

$$K_n = \sup\Big\{l(a), \, a \in \mathfrak{F}_n\Big\},$$

and assume that

$$\limsup_{n \to +\infty} \frac{\log K_n}{n} < 0.$$

Then, φ is finite on an open interval $J \ni -1$.

Theorem 7.4. *Let $x \in \mathbb{R}$ be, such that, $(x+1)\varphi'(x) - \varphi(x) \neq 0$. Then, there exists a Borel probability measure μ_x on $[0,1[$, satisfying*

(1) *for all a, such that, $\mu(a) \neq 0$,*

$$\frac{1}{K}\mu(a)^{x+1}l(a)^{-\varphi(x)} \leq \mu_x(a) \leq K\mu(a)^{x+1}l(a)^{-\varphi(x)}.$$

(2) *$\mu_x(a) = 0$, for $\mu(a) = 0$.*

Proof. Denote as previously, for $\mu(a) \neq 0$,

$$l'(a) = \mu(a)^x l(a)^{-\varphi(x)}.$$

It suffices to prove that

i. $C(x, \varphi(x)) = 0$.

ii. $\lim\limits_{n \to +\infty} \Big(\sup\limits_{a \in \mathbb{A}_n} \mu(a)l'(a)\Big) = 0.$

We start by proving assertion (ii) Let $x < -1$, and J an interval, such that, φ is finite on J. Denote

$$\rho = \inf_{t < -1; t \in J} \frac{\varphi(t)}{t+1}.$$

Whenever $\rho = \dfrac{\varphi(x)}{x+1}$, we get

$$(x+1)\varphi'(x) - \varphi(x) = 0.$$

Otherwise,

$$\rho < \frac{\varphi(z)}{z+1} < \frac{y}{z+1} < \frac{\varphi(x)}{x+1}.$$

We thus get, for $z < -1$, $z \in J$, and $y < \varphi(z)$, $C(z,y) < 0$. Consequently,

$$\sum_{n \geq 1} \left(\sum_{a \in \mathbb{A}_n} \mu(a)^{z+1}l(a)^{-y} \right) < +\infty, \quad \forall a \in \mathbb{A}.$$

Hence,

$$\mu(a)^{z+1}l(a)^{-y} \geq M,$$

which yields that

$$\mu(a) \leq M^{\frac{1}{z+1}}l(a)^{\frac{y}{z+1}}.$$

So, let $a \in \mathbb{A}_n^*$. We immediately obtain

$$\mu(a)^{x+1}l(a)^{-\varphi(x)} \leq M^{\frac{x+1}{z+1}}l(a)^{\frac{y}{z+1}(x+1)-\varphi(x)}.$$

As a result,

$$\sup_{a \in \mathbb{A}_n^*}\left(\mu(a)^{x+1}l(a)^{-\varphi(x)}\right) \leq M^{\frac{x+1}{z+1}}K_n^{\frac{y}{z+1}(x+1)-\varphi(x)} \xrightarrow[n \to +\infty]{} 0.$$

Now, for $x > -1$, and $\rho = \sup\limits_{t>-1}\dfrac{\varphi(t)}{t+1}$, we get $\dfrac{\varphi(x)}{x+1} < \rho$. Finally, for $x = -1$ we obtain $\varphi(-1) \neq 0$, and the same for $\varphi(-1) < 0$. $\qquad\square$

Example 7.1. Take $p = 2$, $\mathbb{A} = \{0,1\}$. Let $\lambda_0, \lambda_1 > 0$, and $\lambda_0 + \lambda_1 = 1$. Consider

$$l(a) = |I_{a_1...a_n}| = \lambda_1...\lambda_{a_n},$$

and

$$\mu(I_{a_1...a_n}) = \frac{1}{2^n}.$$

We claim that

- μ is quasi-Bernoulli.
- l is almost-multiplicative.

Evaluate next $C(x,y)$, and compute φ on $\mathbb{R}$.

Example 7.2. With the same notations as in Example 7.1, in the case of Cantor sets, we consider

$$\mu(\mathcal{R}_i) = \lambda_i, \quad \text{and} \quad 0 \text{ elsewhere,}$$

where $\mathcal{R}_i$ is the interval corresponding to the multi-index i. Remark, in this case, that, for $i = (i_1, i_2, ..., i_n)$, and $j = (j_1, j_2, ..., j_l)$ two multi-indices, one has

$$\mu(\mathcal{R}_{ij}) = \lambda_{ij} = \lambda_i\lambda_j = \mu(\mathcal{R}_i)\mu(\mathcal{R}_j).$$

In this case, μ is a Gibbs measure. Show that the multifractal formalism holds.

7.3 Exercises for Chapter 7

Exercise 1.
Prove Proposition 7.2.

Exercise 2.
Let Ω be a compact subset of $\mathbb{R}^d$, and let ν be a measure defined on subsets of Ω. For a given $s > 0$, let $B(s)$ be an s-covering of Ω. Let ν_j be the measure of a box $B_j \in B(s)$, and let $p_j = \nu_j/\nu(\Omega)$ be the normalized probability of the box B_j. Discarding from $B(s)$ any box for which $p_j = 0$, we assume that $p_j > 0$, for all j, and let $B(s)$ be the number of such boxes. Denote next $\sum_j = \sum\limits_{j=1}^{B(s)}$, and for $q \in \mathbf{R}$,

$$< p^{q-1} > \equiv \sum_j p_j^q.$$

For $q \neq 1$, the generalized dimension of order q is

$$D_q \equiv \frac{1}{q-1} \lim_{s \to 0} \frac{\log < p^{q-1} >}{\log s},$$

assuming that the limit exists. For $q = 1$, the generalized dimension of order 1 is

$$D_1 \equiv \lim_{q \to 1} D_q.$$

1. Show that if the probability measure ν is uniform on Ω, then D_q is independent of q.
2. Prove that D_q (as a function of q) is continuous at $q = 1$.
3. For the Cantor set, prove that if $p_1 = \frac{1}{3}$, then $D_q = 1$ for all q.

Exercise 3.
For the same notations as in Exercise 7.3, define

$$Z_q(s) \equiv \sum_{B_j \in B(s)} [p_j(s)]^q.$$

Let the box sizes u, and v be, such that, $u > v$. For $q \neq 1$, define

$$D_q(u, v) \equiv \frac{\log Z_q(u) - \log Z_q(v)}{(q-1)(\log u - \log v)} = \frac{1}{(q-1)(\log \frac{u}{v})} \log\left(\frac{Z_q(u)}{Z_q(v)}\right),$$

and

$$H(s) \equiv - \sum_{B_j \in B(s)} p_j(s) \log p_j(s).$$

Show that

$$D_1(u,v) \equiv \frac{H(v) - H(u)}{\log(\frac{u}{v})}.$$

Exercise 4.
For an integer $J \geq 2$, let $P(J)$ be the continuous minimization problem

$$\begin{cases} \sum_{j=1}^{J} P_j^q, \\ \text{subject to,} \\ \sum_{j=1}^{J} p_j = 1, \\ p_j \geq 0, \ \forall j. \end{cases}$$

Show that for $q > 1$, the solution of $P(J)$ is $p_j = \frac{1}{J}$ for each j, and that the optimal objective function value is J^{1-q}.

Exercise 5.
Consider a sequence $\{\mathcal{F}_n\}_n$ of finite partitions $\{V_i\}$ of $[0, 1[$, constituted of right-open intervals, such that, $\mathcal{F}_{n+1}$ is a refinement of $\mathcal{F}_n$. For $t \in [0, 1[$, assume that $|\tilde{I}_n(t)| \longrightarrow 0$, as $n \to +\infty$, where $I_n(t)$ is the unique element of $\mathcal{F}_n$ that contains t. Denote finally $\mathcal{F} = \cup_n \mathcal{F}_n$. Let $\xi : \mathcal{F} \longrightarrow \mathbb{R}_+$, and denote

$$H(\xi) = \liminf_{\varepsilon \to 0} \left\{ \sum_j \xi(V_j); \ \{V_j\} \ \text{an} \ \varepsilon - \text{recovering of} \ [0, 1[\right\}.$$

The function ξ is said to be a Frostman function, if $H(\xi) > 0$, and, if 0 is an adherence value of the sequence $\left(\sup_{V \in \mathcal{F}_n} \xi(V) \right)_n$.

Show that if ξ is a Frostman function, there exist a probability measure ν on $[0, 1[$, a constant $M > 0$, and a number $\varepsilon > 0$, such that,

$$\forall V \in \mathcal{F}, \ |V| \leq \varepsilon : \qquad \nu(V) \leq M\xi(V).$$

Exercise 6.
Consider a Cantor type set $\mathcal{C}$ obtained by an iterated function system (I, S_i), $1 = 1, 2$, where $I = 0, 1$. Denote for $n \in \mathbb{N}$, and a multi-index $i = (i_1, ..., i_n) \in \{1, 2\}^n$,

$$I_i = S_{i_1} \circ ... \circ S_{i_n} \circ (I),$$

the corresponding cell in the n-th iteration. Consider next two sequences $(\lambda_i^n)_{n \geq 1}$ in $]0, 1[$, such that,

$$\lambda_i^n + \lambda_i^n = 1, \ \forall n \geq 1.$$

For x, and y real numbers, define

$$\underline{K}(x,y) = \lim_{\varepsilon \to 0} \inf \left\{ \sum_i |\lambda_i|^x |I_i|^{-y}; \ \mathcal{C} \subset \bigcup_i I_i \ \text{and} \ |I_i| \le \varepsilon \right\},$$

$$C_n(x,y) = \frac{1}{n} \log \sum_{|i|=n} |\lambda_i|^x |I_i|^{-y},$$

and

$$C(x,y) = \limsup_{n \to +\infty} C_n(x,y).$$

Let, finally,

$$\varphi(x) = \sup\{y; \ C(x,y) < 0\}.$$

(1) Show that the function φ is non-decreasing, and concave.

(2) Show that $C(x, \varphi(x)) = 0$ for all x.

For the next part, consider, for $j \in \mathbb{Z}$, the set

$$B_j = \{i \ : \ 2^{-j} \le |I_i| < 2.2^{-j}\}.$$

For $i = (i_1, \ldots, i_n)$, denote

$$\lambda_i = \lambda_{i_1}^1 \ldots \lambda_{i_n}^n,$$

$$\alpha_{min} = \liminf_{j \to \infty} \inf_{i \in B_j} \frac{\log \lambda_i}{\log |I_i|},$$

and

$$\alpha_{max} = \limsup_{j \to \infty} \sup_{i \in B_j} \frac{\log \lambda_i}{\log |I_i|}.$$

Assume now that $\underline{K}(p, \varphi(p)) > 0$, $\alpha \in [\alpha_{min}, \alpha_{max}]$, φ is differentiable at p, and $\alpha = \varphi'(p)$.

(3) Show that $d(\alpha) = \alpha p - \varphi(p) = \inf_x (\alpha x - \varphi(x))$.

Exercise 7.

With the same notations as in Exercise 7.3, let $\mathcal{T}_n = \{I_i \ ; \ |i| = n\}$). For a measure μ supported by K, and x, y real numbers, denote

$$\underline{K}_\mu(x,y) = \lim_{\varepsilon \to 0} \inf \left\{ \sum_i \mu(I_i)^x |I_i|^{-y}; \ K \subset \bigcup_i I_i \ \text{and} \ |I_i| \le \varepsilon \right\},$$

and

$$\overline{K}_\mu(x,y) = \lim_{\varepsilon \to 0} \sup \left\{ \sum_i \mu(I_i)^x |I_i|^{-y}; \ K \subset \bigcup_i I_i \ \text{and} \ |I_i| \le \varepsilon \right\}.$$

(1) Show that the set $\{\overline{K}_\mu = 0\}$ is convex.

(2) Show that the function $\overline{K}_\mu$ is decreasing on x, and non-decreasing on y.

(3) Show that, if $\overline{K}_\mu(x, y)$ is finite, then $\overline{K}_\mu(x + t, y - u) = 0$, for $t, u > 0$.

(4) Show that the function $\varphi_\mu : \mathbb{R} \longrightarrow \overline{\mathbb{R}}$, defined by

$$\varphi_\mu(x) = \sup\{y; \quad \overline{K}_\mu(x, y) = 0\},$$

is well defined, concave, and non-decreasing.

(5) Let $\mathcal{F}(\mathbb{R})$ be the set of functions μ defined on $\mathcal{T} = \bigcup_n \mathcal{T}_n$, and $p > 0$.

Show that the set

$$\mathcal{A}_p = \{\mu \in \mathcal{F}(\mathbb{R}) ; \quad \underline{K}_\mu(p, \varphi_\mu(p)) > 0\},$$

is a convex subset of $\mathcal{F}(\mathbb{R})$.

Exercise 8.

Let $k \in \mathbb{N}$ be fixed, and $\lambda_i > 0$, $i = 1, 2, ..., k$, such that, $\sum_{i=1}^{k} \lambda_i < 1$. Denote next

$$F(x) = \log\left(\sum_{i=1}^{k} \lambda_i^x\right).$$

(1) Show that F is real analytic, decreasing, and convex on $\mathbb{R}$.

(2) Find a condition on the λ_i's, $i = 1, 2, ..., k$ for which F is strictly convex.

(3) Describe the asymptotic behaviour of F at $\pm\infty$.

Exercise 9.

Let $g : \mathbb{R}_+ \to \mathbb{R}_+$ be a continuous function, satisfying the following hypothesis,

i. g is non-decreasing.

ii. $g(0) = 0$.

iii. There exist a constant $C > 0$, such that, $g(2r) \leq Cg(r)$, $\forall r \geq 0$.

We denote, for $\varepsilon > 0$, and $E \subset \mathbb{R}^d$,

$$\mathcal{H}_g^\varepsilon(E) = \inf\left\{\sum_i g(diam(U_i)); \ U_i \subset \mathbb{R}^d, \ diam(U_i) < \varepsilon, \ E \subset \bigcup_i U_i\right\},$$

and

$$\mathcal{H}_g(E) = \lim_{\varepsilon \downarrow 0} \mathcal{H}_g^\varepsilon(E),$$

and put *diam* the diameter in the Euclidean distance sense.

1. Show that $\mathcal{H}_g$ is an outer metric measure on $\mathbb{R}^d$.

Let $\mu : \mathcal{B}(\mathbb{R}^d) \to \mathbb{R}_+$ be a non-vanishing finite measure. Let A be a Borel set,, and $K \in]0, +\infty[$ be constant.

2. Assume that, for all $x \in A$, we have

$$\limsup_{r \to 0^+} \frac{\mu(\overline{B}(x,r))}{g(r)} < K.$$

Show that $\mu(A) \leq k\mathcal{H}_g(A)$.

3. Assume that, for all $x \in A$, we have

$$\limsup_{r \to 0^+} \frac{\mu(\overline{B}(x,r))}{g(r)} > K.$$

Show that $\mu(A) \geq C^{-2}k\mathcal{H}_g(A)$.

4. Assume that $\mu(\mathbf{R}^d \setminus A) = 0$, and

$$\int_{\mathbb{R}^d} \int_{\mathbb{R}^d} \frac{\mu(dx)\mu(dy)}{g(|x-y|)} < \infty.$$

Show that $\mathcal{H}_g(A) = \infty$.

5. Assume finally that $\mu(\mathbf{R}^d \setminus A) = 0$, and that, there exists $\alpha > 0$, satisfying

$$\int_{\mathbb{R}^d} \int_{\mathbb{R}^d} \frac{\mu(dx)\mu(dy)}{|x-y|^\alpha} < \infty.$$

Prove that $dim_H(A) \geq \alpha$.

Exercise 10. Let $\beta : \mathbb{R} \to \mathbb{R}$ be a convex function. For a Borel finite measure μ on $\mathbb{R}$, $r > 0$, and $q \in \mathbb{R}$, consider the q-th power moment sums of μ, given by

$$M_r(q) = \sum \mu(A)^q,$$

where the sum is over the r-cubes A, for which $\mu(A) > 0$. Denote next

$$\underline{\beta}(q) = \liminf_{r \to 0} \frac{\log M_r(q)}{-\log r},$$

and

$$\overline{\beta}(q) = \limsup_{r \to 0} \frac{\log M_r(q)}{-\log r}.$$

Denote next $\underline{\beta}_L$, and $\overline{\beta}_L$ the Legendre transforms of $\underline{\beta}$, and $\overline{\beta}$, respectively. Show that

$$\underline{\beta}_L(\alpha) \leq \inf_{-\infty < q < \infty} \{\underline{\beta}(q) + \alpha q\},$$

and

$$\overline{\beta}_L(\alpha) \leq \inf_{-\infty < q < \infty} \{\overline{\beta}(q) + \alpha q\}.$$

Chapter 8

Extensions to Multifractal Cases

The present chapter constitutes a generalization of the concepts studied in the previous ones. All the concepts of fractal measures, such as Hausdorff measure, packing measure, and the fractal dimensions such as, the Hausdorff dimension, the packing dimension, the box, and/or the Bouligand–Minkowski dimension will be reproduced in a general way that permits to cover some large situations, which are not investigated, and/or which may not be investigated by applying the previous concepts.

First rigorous steps in the validation of the multifractal formalism have been conducted in [Brown et al (1992)] for a restrictive class of measures, especially, self-similar. Next, L. Olsen has developed a more general multifractal analysis of measures by including the measure μ itself in the definition of new variants of the Hausdorff measure. A measure depending on two exponents q, t is introduced, by considering sums of the type,

$$\sum_i \mu(B(x_i, r_i))^q (2r_i)^t.$$

The function τ_μ is introduced as a separating value of these sums, and thus permits to show the validity of the multifractal formalism for a large class of multifractal measures by proving that it is always possible to construct a Borel, possibly Gibbs measure ν supported by the α-level sets. More precisely, the measure ν satisfies

$$\nu(B(x, r)) \sim \mu(B(x, r))^q (2r)^t,$$

for all $x \in X_\mu(\alpha)$. To do this, L. Olsen applied the well known large deviation formalism known in probability theory.

Readers may refer to the following references for more details, examples, counter-examples, and also for eventual open problems, and applications, [Batakis and Heurteaux (2002); Ben Mabrouk (2008a,c); Ben Mabrouk and Ben Abdallah (2006); Ben Mabrouk, Ben Abdallah and Dhifaoui (2008); Ben Mabrouk and Aouidi (2011, 2012); Ben Mabrouk, Aouidi and Ben Slimane (2014, 2016); Ben Nasr (1994, 1997); Ben Nasr and Bhouri (1997); Ben Nasr, Bhouri and Heurteaux (2002); David and Semmes (1997); Debussche (1998); Edgar (1998, 2008); Heurteaux (1998); King (1995); Kigami (2001); Lucas (2000); Pesin (1997); Pesin and Climenhaga (2009); Peyriere (1992)].

Recently, the multifractal formalism due to [Brown et al (1992); Olsen (1995)] has been extended to more general cases, such as the mixed multifractal analysis. Readers may refer essentially to [Ben Mabrouk (2008a); Ben Mabrouk, Aouidi and Ben Slimane (2014, 2016); Biswas et al (2012); Das (2005); Falconer et al (1995); Mandelbrot (1995); Menceur, Ben Mabrouk and Betina (2016); Menceur and Ben Mabrouk (2019); Olsen (1996, 2000, 2003, 2004); Olsen et al (2000); Riedi and Scheuring (1997); Ye (2007)].

In all the chapter, we adopt the following conventions, for all real numbers x, and q,

$$\begin{cases} 0^q = \infty & \text{for } q < 0, \\ \forall \ 0^q = 0 \text{ for } q > 0, \\ x^0 = 1 & \text{for }, x \geq 0. \end{cases}$$

8.1 Generalized multifractal versions of the Hausdorff, and packing measures, and dimensions

Let μ be a Borel probability measure on $\mathbb{R}^d$, E a subset of $\mathbb{R}^d$, and $\delta > 0$. For q, t real numbers. Denote

$$\overline{\mathcal{H}}_{\mu,\delta}^{q,t}(E) = \inf \sum_i (\mu(B(x_i, r_i)))^q (2r_i)^t,$$

where the lower bound is taken on all centred δ-coverings of E. As in the construction of the original Hausdorff measure, the quantity $\overline{\mathcal{H}}_{\mu,\delta}^{q,t}(E)$ is a monotone (decreasing) function relative to the variable δ. We thus consider its limit,

$$\overline{\mathcal{H}}_{\mu}^{q,t}(E) = \lim_{\delta \downarrow 0} \overline{\mathcal{H}}_{\mu,\delta}^{q,t}(E) = \sup_{\delta > 0} \overline{\mathcal{H}}_{\mu,\delta}^{q,t}(E).$$

The quantity $\overline{\mathcal{H}}_{\mu}^{q,t}$ is unfortunately not an outer measure on $\mathbb{R}^d$, as it lacks the property of monotony as a function of E, relative to the inclusion order relation. To get an outer measure, we consider

$$\mathcal{H}_{\mu}^{q,t}(E) = \sup_{F \subseteq E} \overline{\mathcal{H}}_{\mu}^{q,t}(F).$$

Next, in a similar way, we write

$$\overline{\mathcal{P}}_{\mu,\delta}^{q,t}(E) = \sup \sum_i (\mu(B(x_i,r_i)))^q (2r_i)^t,$$

where the upper bound is taken on all δ-packings of E (composed of balls centered in E, and with diameters less than or equal to δ). As previously, we obtain a decreasing function relative to the variable δ. We thus consider its limit

$$\overline{\mathcal{P}}_{\mu}^{q,t}(E) = \lim_{\delta \downarrow 0} \overline{\mathcal{P}}_{\mu,\delta}^{q,t}(E) = \inf_{\delta > 0} \overline{\mathcal{P}}_{\mu,\delta}^{q,t}(E),$$

which is also not an outer measure, as it lacks the property of sub-additivity. To get an outer measure, we put

$$\mathcal{P}_{\mu}^{q,t}(E) = \inf_{E \subseteq \cup_i E_i} \sum_i \overline{\mathcal{P}}_{\mu}^{q,t}(E_i).$$

Finally, we take as a convention (which in fact may be proved),

$$\mathcal{H}_{\mu,\delta}^{q,t}(\emptyset) = \mathcal{P}_{\mu,\delta}^{q,t}(\emptyset) = 0,$$

or equivalently,

$$\mathcal{H}_{\mu}^{q,t}(\emptyset) = \mathcal{P}_{\mu}^{q,t}(\emptyset) = 0.$$

Proposition 8.1. $\mathcal{H}_{\mu}^{q,t}$, and $\mathcal{P}_{\mu}^{q,t}$ are metric outer measures on $\mathbb{R}^d$.

Proof. We will split the proof into the following steps.

(1) $\mathcal{H}_{\mu}^{q,t}$ is an outer measure.
(2) $\mathcal{H}_{\mu}^{q,t}$ is metric.
(3) $\mathcal{P}_{\mu}^{q,t}$ is an outer measure.
(4) $\mathcal{P}_{\mu}^{q,t}$ is metric.

(1) Recall that $\mathcal{H}_{\mu}^{q,t}$ is an outer measure, if it satisfies the following three assertions,

   (1.i) $\mathcal{H}_{\mu}^{q,t}(\emptyset) = 0$.
   (1.ii) $\mathcal{H}_{\mu}^{q,t}$ is monotone, in the sense that, for any subsets $E \subseteq F \subset \mathbb{R}^d$, we have

$$\mathcal{H}_{\mu}^{q,t}(E) \leq \mathcal{H}_{\mu}^{q,t}(F).$$

   (1.iii) $\mathcal{H}_{\mu}^{q,t}$ is sub-additive, in the sense that, for any sequence $(A_n)_n$ of subsets of $\mathbb{R}^d$, we have

$$\mathcal{H}_{\mu}^{q,t}(\bigcup_n A_n) \leq \sum_n \mathcal{H}_{\mu}^{q,t}(A_n).$$

So, let us develop each point.

(1.i) This is true by the construction convention on $\overline{\mathcal{H}}_\mu^{q,t}$, and $\mathcal{H}_\mu^{q,t}$.

(1.ii) Let $E \subseteq F$ be two subsets of $\mathbb{R}^d$. We have

$$\mathcal{H}_\mu^{q,t}(E) = \sup_{A \subseteq E} \overline{\mathcal{H}}_\mu^{q,t}(A) \le \sup_{A \subseteq F} \overline{\mathcal{H}}_\mu^{q,t}(A) = \mathcal{H}_\mu^{q,t}(F).$$

Consequently,

$$\mathcal{H}_\mu^{q,t}(E) \le \mathcal{H}_\mu^{q,t}(F).$$

(1.iii) Consider firstly a sequence $(E_n)_n$ of subsets in $\mathbb{R}^d$, such that,

$$\sum_n \mathcal{H}_\mu^{q,t}(E_n) < \infty.$$

Let $\delta, \epsilon > 0$, and $(B(x_{ni}, r_{ni}))_i$ be a centred δ-covering of E_n, for which, we have

$$\sum_i \mu(B(x_{ni}, r_{ni}))^q (2r_{ni})^t \le \overline{\mathcal{H}}_{\mu,\delta}^{q,t}(E_n) + \frac{\epsilon}{2^n}.$$

The collection $(B(x_{ni}, r_{ni}))_{n,i}$ is then a centred δ-covering of $\bigcup_n E_n$. As a result,

$$\begin{aligned}
\overline{\mathcal{H}}_{\mu,\delta}^{q,t}\left(\bigcup_n E_n\right) &\le \sum_n \sum_i (\mu(B(x_{ni}, r_{ni})))^q (2r_{ni})^t \\
&\le \sum_n \left(\overline{\mathcal{H}}_{\mu,\delta}^{q,t}(E_n) + \frac{\epsilon}{2^n}\right) \\
&\le \sum_n \left(\overline{\mathcal{H}}_\mu^{q,t}(E_n) + \frac{\epsilon}{2^n}\right) \\
&\le \sum_n \mathcal{H}_\mu^{q,t}(E_n) + \epsilon.
\end{aligned}$$

By letting δ, and ϵ going to 0, we obtain

$$\overline{\mathcal{H}}_\mu^{q,t}\left(\bigcup_n E_n\right) \le \sum_n \mathcal{H}_\mu^{q,t}(E_n).$$

Let now $(A_n)_n$ be a covering of F by subsets of $\mathbb{R}^d$. We have

$$\begin{aligned}
\overline{\mathcal{H}}_\mu^{q,t}(F) &= \overline{\mathcal{H}}_\mu^{q,t}\left(\bigcup_n (A_n F)\right) \\
&\le \sum_n \mathcal{H}_\mu^{q,t}(A_n \cap F) \\
&\le \sum_n \mathcal{H}_\mu^{q,t}(A_n).
\end{aligned}$$

Taking the sup on $F \subseteq \bigcup_n A_n$, we get

$$\mathcal{H}_\mu^{q,t}(\bigcup_n A_n) \leq \sum_n \mathcal{H}_\mu^{q,t}(A_n).$$

(2) We now prove that $\mathcal{H}_\mu^{q,t}$ is a metric measure on $\mathbb{R}^d$. Indeed, consider two subsets E_1, E_2 of $\mathbb{R}^d$, such that,

$$d(E_1, E_2) > 0, \quad \text{and} \quad \mathcal{H}_\mu^{q,t}(E_1 \cup E_2) < \infty.$$

Let also

$$0 < \delta < d(E_1, E_2), \quad \epsilon > 0, \quad F_1 \subseteq E_1, \quad F_2 \subseteq E_2,$$

and $(B(x_i, r_i))_i$ be a centred δ-covering of $F_1 \cup F_2$, such that,

$$\overline{\mathcal{H}}_{\mu,\delta}^{q,t}(F_1 \cup F_2) \leq \sum_i (\mu(B(x_i, r_i)))^q (2r_i)^t \leq \overline{\mathcal{H}}_{\mu,\delta}^{q,t}(F_1 \cup F_2) + \epsilon.$$

Consider next

$$I = \{\, i; \ B(x_i, r_i) \cap F_1 \neq \emptyset \,\} \quad \text{and} \quad J = \{\, i; \ B(x_i, r_i) \cap F_2 \neq \emptyset \,\}.$$

The countable sets $(B(x_i, r_i))_{i \in I}$, and $(B(x_i, r_i))_{i \in J}$ are centred δ-coverings of F_1, and F_2, respectively. Hence,

$$\begin{aligned}
\overline{\mathcal{H}}_{\mu,\delta}^{q,t}(F_1) + \overline{\mathcal{H}}_{\mu,\delta}^{q,t}(F_2) &\leq \sum_{i \in I} (\mu(B(x_i, r_i)))^q (2r_i)^t + \sum_{i \in J} \mu(B(x_i, r_i))^q (2r_i)^t \\
&= \sum_i (\mu(B(x_i, r_i)))^q (2r_i)^t \\
&\leq \overline{\mathcal{H}}_{\mu,\delta}^{q,t}(F_1 \cup F_2) + \epsilon.
\end{aligned}$$

which leads to

$$\overline{\mathcal{H}}_\mu^{q,t}(F_1) + \overline{\mathcal{H}}_\mu^{q,t}(F_2) \leq \overline{\mathcal{H}}_\mu^{q,t}(F_1 \cup F_2) + \epsilon \leq \mathcal{H}_\mu^{q,t}(E_1 \cup E_2) + \epsilon.$$

By letting $\epsilon \downarrow 0$, and taking next the sup on $F_1 \subseteq E_1$, and $F_2 \subseteq E_2$, we obtain

$$\mathcal{H}_\mu^{q,t}(E_1 \cup E_2) \geq \mathcal{H}_\mu^{q,t}(E_1) + \mathcal{H}_\mu^{q,t}(E_2).$$

The other inequality comes from the subadditivity of $\mathcal{H}_\mu^{q,t}$.

(3) To prove that $\mathcal{P}_\mu^{q,t}$ is an outer measure, we have to show the analogues of the three assertions (1.i), (1.ii), and (1.iii) above for $\mathcal{P}_\mu^{q,t}$ instead of $\mathcal{H}_\mu^{q,t}$. We thus designate by (3.i), (3.ii), and (3.iii) the new analogues.

(3.i) This is obvious from the construction of $\mathcal{P}_\mu^{q,t}$.

(3.ii) Let $E \subseteq F$ be two subsets of $\mathbb{R}^d$. We have

$$\mathcal{P}_\mu^{q,t}(E) = \inf_{E \subseteq \bigcup_i E_i} \sum_i \overline{\mathcal{P}}_\mu^{q,t}(E_i) \leq \inf_{F \subseteq \bigcup_i E_i} \sum_i \overline{\mathcal{P}}_\mu^{q,t}(E_i) = \mathcal{P}_\mu^{q,t}(F).$$

(3.iii) Let now $(A_n)_n$ be a sequence of subsets of $\mathbb{R}^d$, $\epsilon > 0$, and $(E_{ni})_i$ be a covering of A_n, such that,

$$\sum_i \overline{\mathcal{P}}_\mu^{q,t}(E_{ni}) \leq \mathcal{P}_\mu^{q,t}(A_n) + \frac{\epsilon}{2^n}.$$

It result, for all $\epsilon > 0$, that

$$\mathcal{P}_\mu^{q,t}(\bigcup_n A_n) \leq \sum_n \sum_i \overline{\mathcal{P}}_\mu^{q,t}(E_{ni}) \leq \sum_n \mathcal{P}_\mu^{q,t}(A_n) + \epsilon,$$

which leads to

$$\mathcal{P}_\mu^{q,t}(\bigcup_n A_n) \leq \sum_n \mathcal{P}_\mu^{q,t}(A_n).$$

(4) We now prove that $\mathcal{P}_\mu^{q,t}$ is a metric measure on $\mathbb{R}^d$. So, let A, and B be two subsets of $\mathbb{R}^d$, such that, $d(A,B) > 0$. We shall show that

$$\mathcal{P}_\mu^{q,t}(A \cup B) = \mathcal{P}_\mu^{q,t}(A) + \mathcal{P}_\mu^{q,t}(B).$$

Indeed, since $\mathcal{P}_\mu^{q,t}$ is an outer measure, it suffices to prove that

$$\mathcal{P}_\mu^{q,t}(A \cup B) \geq \mathcal{P}_\mu^{q,t}(A) + \mathcal{P}_\mu^{q,t}(B).$$

Without loss of generality, we may assume that $\mathcal{P}_\mu^{q,t}(A \cup B) < \infty$. For $\epsilon > 0$, let $(E_i)_i$ be a covering of $A \cup B$, such that,

$$\sum_i \overline{\mathcal{P}}_\mu^{q,t}(E_i) \leq \mathcal{P}_\mu^{q,t}(A \cup B) + \epsilon.$$

Consider $F_i = A \cap E_i$, and $H_i = B \cap E_i$. Then, the collections $(F_i)_i$, and $(H_i)_i$ are two coverings of A, and B, respectively. Moreover, $F_i \cap H_j = \emptyset$ for all i, and j. Besides

$$\mathcal{P}_\mu^{q,t}(A) + \mathcal{P}_\mu^{q,t}(B) \leq \sum_i (\overline{\mathcal{P}}_\mu^{q,t}(F_i) + \overline{\mathcal{P}}_\mu^{q,t}(H_i)).$$

Now, recall that $d(A,B) > 0$, which implies that

$$\overline{\mathcal{P}}_\mu^{q,t}(A \cup B) = \overline{\mathcal{P}}_\mu^{q,t}(A) + \overline{\mathcal{P}}_\mu^{q,t}(B).$$

It follows that

$$\mathcal{P}_\mu^{q,t}(A) + \mathcal{P}_\mu^{q,t}(B) \leq \sum_i \overline{\mathcal{P}}_\mu^{q,t}(E_i) \leq \mathcal{P}_\mu^{q,t}(A \cup B) + \epsilon.$$

$\square$

The following proposition, based on the well known Besicovitch covering theorem, gives a comparison between the quantities $\mathcal{H}_\mu^{q,t}$, $\mathcal{P}_\mu^{q,t}$, and $\overline{\mathcal{P}}_\mu^{q,t}$.

Proposition 8.2. *There exists a constant $C > 0$, such that, for all subset $E \subset \mathbb{R}^d$, and for all $q, t \in \mathbb{R}$, we have*

$$\mathcal{H}_\mu^{q,t} \leq C\mathcal{P}_\mu^{q,t} \leq C\overline{\mathcal{P}}_\mu^{q,t}.$$

Proof. The right inequality holds for all constant $C > 0$. We shall thus prove the left one. Let $F \subseteq \mathbb{R}^d$, $\delta > 0$, and

$$\mathcal{V} = \{B(x, \tfrac{1}{2}); \ x \in F\}.$$

Let also $\big((B_{ij})_j\big)_{1 \leq i \leq \xi}$ be the ξ families of $\mathcal{V}$ defined in Besicovitch's covering Theorem, such that, $(B_{ij})_{i,j}$ is a centred δ-covering of F, and $(B_{ij})_j$ is a centred δ-packing of F, for all $i \in \mathbb{N}$. We have

$$\overline{\mathcal{H}}_{\mu,\delta}^{q,t}(F) \leq \sum_{i=1}^{\xi}\sum_{j} \big(\mu(B_{ij})\big)^q (2r_{ij})^t \leq \sum_{i=1}^{\xi} \overline{\mathcal{P}}_{\mu,\delta}^{q,t}(F) = \xi\overline{\mathcal{P}}_{\mu,\delta}^{q,t}(F).$$

As a result,

$$\overline{\mathcal{H}}_\mu^{q,t}(F) \leq \xi\overline{\mathcal{P}}_\mu^{q,t}(F).$$

Consequently, if $E \subseteq \bigcup_i E_i$, we get

$$\mathcal{H}_\mu^{q,t}(E) = \mathcal{H}_\mu^{q,t}\big(\bigcup_i (E_i \cap E)\big)$$

$$\leq \sum_i \mathcal{H}_\mu^{q,t}(E_i \cap E)$$

$$\leq \sum_i \sup_{F \subseteq E_i \cap E} \overline{\mathcal{H}}_\mu^{q,t}(F)$$

$$\leq \xi \sum_i \sup_{F \subseteq E_i \cap E} \overline{\mathcal{P}}_\mu^{q,t}(F)$$

$$\leq \xi \sum_i \overline{\mathcal{P}}_\mu^{q,t}(E_i),$$

which gives the inequality stated, with $C = \xi$. □

Proposition 8.3. *Let E be a subset of $\mathbb{R}^d$.*

(1) *There exists a unique element $\alpha_0 \in \overline{\mathbb{R}}$, such that,*

$$\mathcal{H}_\mu^{q,t}(E) = \begin{cases} \infty & \text{for } t < \alpha_0, \\ 0 & \text{for } t > \alpha_0. \end{cases}$$

(2) *There exists a unique element $\beta_0 \in \overline{\mathbb{R}}$, such that,*

$$\mathcal{P}_\mu^{q,t}(E) = \begin{cases} \infty & \text{for } t < \beta_0, \\ 0 & \text{for } t > \beta_0. \end{cases}$$

(3) *There exists a unique element $\gamma_0 \in \overline{\mathbb{R}}$, such that,*

$$\overline{\mathcal{P}}_\mu^{q,t}(E) = \begin{cases} \infty & \text{for } t < \gamma_0, \\ 0 & \text{for } t > \gamma_0. \end{cases}$$

Proof. (1) We claim firstly that

$$\forall t \in \mathbb{R}, \quad \mathcal{H}_\mu^{q,t}(E) < \infty \implies \mathcal{H}_\mu^{q,t'}(E) = 0, \quad \forall t' > t. \tag{8.1}$$

Indeed, let $\delta > 0$, $F \subseteq E$, and $(B(x_i, r_i))_i$ be a centred δ-covering of F. We have

$$\overline{\mathcal{H}}_{\mu,\delta}^{q,t'}(F) \le \sum_i (\mu(B(x_i, r_i)))^q (2r_i)^{t'}$$
$$\le \delta^{t'-t} \sum_i (\mu(B(x_i, r_i)))^q (2r_i)^t.$$

Therefore,

$$\overline{H}_{\mu,\delta}^{q,t'}(F) \le \delta^{t'-t} \overline{H}_{\mu,\delta}^{q,t'}(F),$$

which leads to

$$\overline{H}_\mu^{q,t'}(F) = 0, \quad \forall F \subseteq E.$$

Consequently,

$$\mathcal{H}_\mu^{q,t'}(E) = 0.$$

We thus put

$$\alpha_0 = \inf\{ t \in \mathbb{R} \ \mathcal{H}_\mu^{q,t'}(E) = 0 \}.$$

(2) Let $t \in \mathbb{R}$ be such that, $\mathcal{P}_\mu^{q,t}(E) < \infty$. There exists a sequence $(E_i)_i$ of subsets of $\mathbb{R}^d$ satisfying

$$E \subseteq \bigcup_i E_i, \quad \text{and} \quad \overline{P}_\mu^{q,t}(E_i) < \infty, \quad \forall i.$$

Consider next, $t' > t$, $\delta > 0$, and $(B(x_{ni}, r_{ni}))_n$ a centred δ-packing of E_i. We have

$$\sum_n (\mu(B(x_{ni}, rni)))^q (2r_{ni})^{t'} \le \delta^{t'-t} \sum_n (\mu(B(x_{ni}, rni)))^q (2r_{ni})^t.$$

Therefore,

$$\overline{P}_{\mu,\delta}^{q,t'}(E_i) \le \delta^{t'-t}\overline{P}_{\mu,\delta}^{q,t}(E_i),$$

which leads to

$$\overline{P}_\mu^{q,t'}(E_i) = 0, \; \forall \, i.$$

Consequently,

$$\mathcal{P}_\mu^{q,t'}(E) = 0.$$

We thus put, as previously,

$$\beta_0(E) = \inf\{t \in \mathbb{R}; \mathcal{P}_\mu^{q,t}(E) = 0\}.$$

(3) Let $t \in \mathbb{R}$ be such that, $\overline{P}_\mu^{q,t}(E) < \infty$. For all centred δ-packing $(B(x_i, r_i))_i$ of E, and all $t' > t$, we have

$$\sum_i (\mu(B(x_i, r_i)))^q (2r_i)^{t'} \le \delta^{t'-t} \sum_i (\mu(B(x_i, r_i)))^q (2r_i)^t.$$

Consequently,

$$\overline{P}_\mu^{q,t'}(E) = 0.$$

We thus set

$$\gamma_0 = \inf\{t \in \mathbb{R}; \mathcal{P}_\mu^{q,t}(E) = 0\}.$$

□

Definition 8.1. The cut-off values α_0, β_0, and σ_0 are called, respectively, the multifractal generalizations of the Hausdorff dimension, packing, and the logarithmic index of E. We denote them, respectively, $dim_\mu^q(E)$, $Dim_\mu^q(E)$, and $\Delta_\mu^q(E)$.

The following proposition resumes some usual, and elementary characteristics of these dimensions.

Proposition 8.4.

(1) dim_μ^q, and Dim_μ^q are monotone. i.e., for $E \subseteq F$, we have

$$i) \qquad dim_\mu^q(E) \le dim_\mu^q(F),$$

and

$$ii) \qquad Dim_\mu^q(E) \le Dim_\mu^q(F).$$

(2) dim_μ^q, and Dim_μ^q are σ-stables. i.e

$$i) \qquad dim_\mu^q(\bigcup_n E_n) = \sup_n dim_\mu^q(E_n),$$

and

$$ii) \qquad Dim_\mu^q(\bigcup_n E_n) = \sup_n Dim_\mu^q(E_n).$$

Proof. (1.i) The monotony of dim_μ^q. Let $E \subseteq F$ be two subsets of $\mathbb{R}^d$, such that, $dim_\mu^q(F) < \infty$. We have

$$\mathcal{H}_\mu^{q,t}(F) = 0, \quad \forall\, t > dim_\mu^q(F).$$

It results, from the monotony of $\mathcal{H}_\mu^{q,t}$, that

$$\mathcal{H}_\mu^{q,t}(E) = 0, \quad \forall\, t > dim_\mu^q(F).$$

Hence,

$$dim_\mu^q(E) \le t, \quad \forall\, t > dim_\mu^q(F).$$

Consequently,

$$dim_\mu^q(E) \le dim_\mu^q(F).$$

(1.ii) The monotony of Dim_μ^q. Let $E \subseteq F$ be two subsets of $\mathbb{R}^d$, such that, $Dim_\mu^q(F) < \infty$. Hence,

$$\mathcal{P}_\mu^{q,t}(F) = 0, \quad \forall\, t > Dim_\mu^q(F).$$

As $\mathcal{P}_\mu^{q,t}$ is an outer measure, we deduce that

$$\mathcal{P}_\mu^{q,t}(E) = 0, \quad \forall\, t > Dim_\mu^q(F).$$

Thus,

$$Dim_\mu^q(E) \le t, \quad \forall\, t > Dim_\mu^q(F).$$

Therefore,

$$Dim_\mu^q(E) \le Dim_\mu^q(F).$$

(2.i) The σ-stability of dim_μ^q. Let $(E_n)_n$ be a sequence of subsets $\mathbb{R}^d$, such that,

$$dim_\mu^q(\bigcup_n E_n) < \infty.$$

We have,

$$E_n \subseteq \bigcup_m E_m, \quad \forall\, n.$$

We deduce, from assertion **1.i.** above, that, for all n,

$$dim_\mu^q(E_n) \le dim_\mu^q(\bigcup_n E_n).$$

Therefore,

$$\sup_n dim_\mu^q(E_n) \le dim_\mu^q(\bigcup_n E_n).$$

Let now, $t > \sup_n dim_\mu^q(E_n)$. We have

$$\mathcal{H}_\mu^{q,t}(E_n) = 0, \quad \forall\, n.$$

It follows, from the sub-additivity of $\mathcal{H}_\mu^{q,t}$, that

$$\mathcal{H}_\mu^{q,t}(\bigcup_n E_n) = 0, \quad \forall\, t > \sup_n dim_\mu^q(E_n).$$

Consequently,

$$dim_\mu^q(\bigcup_n E_n) \le t, \quad \forall\, t > \sup_n dim_\mu^q(E_n).$$

As a result,

$$dim_\mu^q(\bigcup_n E_n) \le \sup_n dim_\mu^q(E_n).$$

(2.ii) The σ-stability of Dim_μ^q. Let $(E_n)_n$ be a sequence of subsets $\mathbb{R}^d$. Observing that

$$E_n \subseteq \bigcup_m E_m, \quad \forall\, n,$$

we get

$$\sup_n Dim_\mu^q(E_n) \le Dim_\mu^q(\bigcup_n E_n).$$

Next, assume that $\sup_n Dim_\mu^q(E_n) < \infty$, and consider $t > \sup_n Dim_\mu^q(E_n)$. We obtain,

$$\mathcal{P}_\mu^{q,t}(E_n) = 0, \forall n.$$

It follows, from the sub-additivity of $\mathcal{P}_\mu^{q,t}$, that

$$\mathcal{P}_\mu^{q,t}(\bigcup_n E_n) = 0, \quad \forall\, t > \sup_n Dim_\mu^q(E_n).$$

Consequently,

$$Dim_\mu^q(\bigcup_n E_n) \le t, \quad \forall\, t > \sup_n Dim_\mu^q(E_n).$$

Hence,

$$Dim_\mu^q(\bigcup_n E_n) \le \sup_n Dim_\mu^q(E_n).$$

$\square$

In the rest of the chapter, we will denote

$$b_{\mu,E}(q) = dim_\mu^q(E), \quad B_{\mu,E}(q) = Dim_\mu^q(E), \quad \text{and} \quad \Lambda_{\mu,E}(q) = \Delta_\mu^q(E).$$

In the case where E is the support of μ, we will simply denote

$$b_\mu(q) = dim_\mu^q(E), \quad B_\mu(q) = Dim_\mu^q(E), \quad \text{and} \quad \Lambda_\mu(q) = \Delta_\mu^q(E).$$

Proposition 8.5. *The following assertions hold.*

(1) $b_{\mu,E}$, $B_{\mu,E}$, *and* $\Lambda_{\mu,E}$ *are decreasing.*
(2) $B_{\mu,E}$, *and* $\Lambda_{\mu,E}$ *are convex.*
(3) $0 \leq b_\mu(q) \leq B_\mu(q) \leq \Lambda_\mu(q)$, $\forall q < 1$.
(4) $b_\mu(1) \leq B_\mu(1) \leq \Lambda_\mu(1) = 0$.
(5) $b_\mu(q) \leq B_\mu(q) \leq \Lambda_\mu(q) \leq 0$, $\forall q > 1$.

Proof. (1) The monotony of $b_{\mu,E}$. Let $p \geq q$ be two real numbers, $F \subseteq E$, $\delta > 0$, and $(B(x_i, r_i))_i$ a centred δ-covering of F. For all $t \in \mathbb{R}$, we have

$$\sum_i (\mu(B(x_i, r_i)))^p (2r_i)^t \leq \sum_i (\mu(B(x_i, r_i)))^q (2r_i)^t.$$

Hence,

$$\overline{\mathcal{H}}_{\mu,\delta}^{p,t}(F) \leq \overline{\mathcal{H}}_{\mu,\delta}^{q,t}(F).$$

By taking the limit as $\delta \downarrow 0$, we obtain

$$\overline{\mathcal{H}}_\mu^{p,t}(F) \leq \overline{\mathcal{H}}_\mu^{q,t}(F).$$

This leads to

$$\mathcal{H}_\mu^{p,t}(E) \leq \mathcal{H}_\mu^{q,t}(E).$$

Consequently,

$$\mathcal{H}_\mu^{p,t}(E) = 0, \quad \forall\, t > b_{\mu,E}(q).$$

Therefore,

$$b_{\mu,E}(p) < t, \quad \forall\, t > b_{\mu,E}(q).$$

Hence,

$$b_{\mu,E}(p) \leq b_{\mu,E}(q).$$

We shall now prove the monotony of $B_{\mu,E}$. Let $(E_i)_i$ be a covering of E by a sequence of subsets of $\mathbb{R}^d$. If holds, for $p \geq q$, that

$$\sum_i \overline{\mathcal{P}}_\mu^{p,t}(E_i) \leq \sum_i \overline{\mathcal{P}}_\mu^{q,t}(E_i).$$

Therefore,

$$\mathcal{P}_\mu^{p,t}(E) \leq \mathcal{P}_\mu^{q,t}(E).$$

It results that

$$\mathcal{P}_\mu^{p,t}(E) = 0, \quad \forall\, t > B_{\mu,E}(q).$$

Consequently,

$$B_{\mu,E}(p) < t, \quad \forall\, t > B_{\mu,E}(q).$$

Thus,

$$B_{\mu,E}(p) \leq B_{\mu,E}(q).$$

We now prove the monotony of $\Lambda_{\mu,E}$. Let $p \geq q$ be two real numbers, $\delta > 0$, and $(B(x_i, r_i))_i$ be a centred δ-packing of E. We have

$$\mu(B(x_i, r_i))^p (2r_i)^t \leq \mu(B(x_i, r_i))^q (2r_i)^t, \; \forall\, t \in \mathbb{R}.$$

It results that

$$\overline{P}_{\mu,\delta}^{p,t}(E) \leq \overline{P}_{\mu,\delta}^{q,t}(E).$$

By taking the limit on $\delta \downarrow 0$, we obtain

$$\overline{P}_\mu^{p,t}(E) \leq \overline{P}_\mu^{q,t}(E).$$

It follows that,

$$\overline{P}_\mu^{p,t}(E) = 0, \quad \forall\, t > \Lambda_{\mu,E}(q).$$

Consequently,

$$\Lambda_{\mu,E}(p) < t, \quad \forall\, t > \Lambda_{\mu,E}(q).$$

Hence,

$$\Lambda_{\mu,E}(p) \leq \Lambda_{\mu,E}(q).$$

(2) The convexity of $B_{\mu,E}$. Consider $p, q \in \mathbb{R}$, $\alpha \in [0,1]$, $\epsilon > 0$, $t = B_{\mu,E}(q)$, and $s = B_{\mu,E}(p)$. We have

$$\mathcal{P}_\mu^{q,t+\epsilon}(E) = \mathcal{P}_\mu^{p,s+\epsilon}(E) = 0.$$

Consequently, There exits two coverings $(H_i)_i$, and $(K_i)_i$ of E, such that,

$$\sum_i \overline{P}_\mu^{q,t+\epsilon}(H_i) \leq 1,$$

and

$$\sum_i \overline{P}_\mu^{p,s+\epsilon}(K_i) \leq 1.$$

Consider, for each $n \in \mathbb{N}$, the set

$$E_n = \bigcup_{1 \leq i,j \leq n} (H_i \cap K_j).$$

As $\mathcal{P}_\mu^{q,t}$ is an outer measure, we obtain

$$\mathcal{P}_\mu^{\alpha q + (1-\alpha)p, \alpha t + (1-\alpha)s + \epsilon}(E_n)$$

$$\leq \sum_{i,j=1}^n \mathcal{P}_\mu^{\alpha q + (1-\alpha)p, \alpha t + (1-\alpha)s + \epsilon}(H_i \cap K_j)$$

$$\leq \sum_{i,j=1}^n \overline{\mathcal{P}}_\mu^{\alpha q + (1-\alpha)p, \alpha t + (1-\alpha)s + \epsilon}(H_i \cap K_j)$$

$$\leq \left(\sum_{i,j=1}^n \overline{\mathcal{P}}_\mu^{q,t+\epsilon}(H_i \cap K_j) \right)^\alpha \left(\sum_{i,j=1}^n \overline{\mathcal{P}}_\mu^{p,s+\epsilon}(H_i \cap K_j) \right)^{1-\alpha}$$

$$\leq n^\alpha n^{1-\alpha} = n < \infty.$$

Consequently,

$$B_{\mu,E_n}(\alpha q + (1-\alpha)p) \leq \alpha t + (1-\alpha)s + \epsilon, \quad \forall\, \epsilon > 0.$$

As a result,

$$B_{\mu,E}(\alpha q + (1-\alpha)p) \leq \alpha B_{\mu,E}(q) + (1-\alpha)B_{\mu,E}(p).$$

We now show the convexity of $\Lambda_{\mu,E}$, let p, q be two real numbers, $\alpha \in [0,1]$,

$$s > \Lambda_{\mu,E}(p), \quad \text{and} \quad t > \Lambda_{\mu,E}(q).$$

Let also $\delta > 0$, and $(B_i = B(x_i, r_i))_i$ be a centred δ-packing of E. From Hölder's inequality, we get

$$\sum_i (\mu(B_i))^{\alpha q + (1-\alpha)p}(2r_i)^{\alpha t + (1-\alpha)s}$$

$$\leq \left(\sum_i (\mu(B_i))^q (2r_i)^t \right)^\alpha \left(\sum_i (\mu(B_i))^p (2r_i)^s \right)^{1-\alpha},$$

which leads to

$$\overline{\mathcal{P}}_{\mu,\delta}^{\alpha q + (1-\alpha)p, \alpha t + (1-\alpha)s}(E) \leq \left(\overline{\mathcal{P}}_{\mu,\delta}^{q,t}(E) \right)^\alpha \left(\overline{\mathcal{P}}_{\mu,\delta}^{p,s}(E) \right)^{1-\alpha}.$$

By taking the limit as $\delta \downarrow 0$, we obtain

$$\overline{\mathcal{P}}_\mu^{\alpha q + (1-\alpha)p, \alpha t + (1-\alpha)s}(E) \leq \left(\overline{\mathcal{P}}_\mu^{q,t}(E) \right)^\alpha \left(\overline{\mathcal{P}}_\mu^{p,s}(E) \right)^{1-\alpha}.$$

Consequently,

$$\overline{\mathcal{P}}_\mu^{\alpha q + (1-\alpha)p, \alpha t + (1-\alpha)s}(E) = 0.$$

Now, as s and t are arbitrary chosen, with

$$s > \Lambda_{\mu,E}(p), \quad \text{and} \quad t > \Lambda_{\mu,E}(q),$$

we deduce that

$$\Lambda_{\mu,E}(\alpha q + (1-\alpha)p) \leq \alpha \Lambda_{\mu,E}(q) + (1-\alpha)\Lambda_{\mu,E}(p).$$

(3) It follows, from Proposition 8.2, that

$$b_\mu(q) \leq B_\mu(q) \leq \Lambda_\mu(q), \quad \forall p \in \mathbb{R}. \tag{8.2}$$

(4) From assertion **(3)**, it suffices to show that

$$b_\mu(1) \geq 0, \quad \text{and} \quad \Lambda_\mu(1) \leq 0.$$

Indeed, if $t < 0$, $0 < \delta < \frac{1}{2}$, and $\big(B(x_i, r_i)\big)_i$ is a centred δ-covering of $Support(\mu)$, we get

$$\sum_i \mu(B(x_i, r_i))(2r_i)^t \geq 1,$$

which leads to

$$\overline{\mathcal{H}}_{\mu,\delta}^{1,t}(Support(\mu)) \geq 1, \quad \forall\, t < 0.$$

Consequently,

$$b_\mu(1) \geq t, \quad \forall\, t < 0 \;\Rightarrow\; b_\mu(1) \geq 0.$$

Consider now $t > 0$, $0 < \delta < \frac{1}{2}$, and $\big(B(x_i, r_i)\big)_i$ a centred δ-packing of $Support(\mu)$. We have

$$\overline{\mathcal{P}}_{\mu,\delta}^{1,t}(Support(\mu)) \leq \sum_i \mu(B(x_i, r_i))(2r_i)^t \leq 1.$$

Thus,

$$\overline{\mathcal{P}}_{\mu,\delta}^{1,t}(Support(\mu)) \leq 1, \quad \forall\, t < 0.$$

Therefore,

$$\Lambda_\mu(1) \leq t, \quad \forall\, t < 0,$$

which yields, finally, that

$$\Lambda_\mu(1) \leq 0.$$

(5) is similar to assertion **(3)**. $\qquad\qquad\qquad\qquad\qquad\qquad\square$

Remark 8.1.

(a) $b_{\mu,E}(0) = dim\, E$.
(b) $B_{\mu,E}(0) = Dim\, E$.
(c) $\Lambda_{\mu,E}(0) = \Delta(E)$.
(d) $B_{\mu,E}$, and $\Lambda_{\mu,E}$ are finite on $[0, +\infty[$.

8.2 Generalized Bouligand–Minkowski dimension

We consider a Borel probability measure μ on $\mathbb{R}^d$. For $a > 1$, and E a subset of $Support(\mu)$, we write

$$T_a(E) = \limsup_{r \downarrow 0} \left(\sup_{x \in E} \frac{\mu(B(x, ar))}{\mu(B(x, r))} \right),$$

and for $x \in Support(\mu)$, we write $T_a(x) = T_a(\{x\})$.

Lemma 8.1. *The two following assertions are equivalent.*

(1) $\forall a > 1$, $T_a(E) < \infty$.
(2) *There exists $a > 1$, such that, $T_a(E) < \infty$.*

Proof. The implication **1)** $\Longrightarrow$ **2)** is obvious. We shall prove the opposite sense. Let $a > 1$ be, such that, $T_a(E) < \infty$, and $b > 1$. There exists $n \in \mathbb{N}$, such that, $b \leq a^n$. Thus,

$$T_b(E) = \limsup_{r \downarrow 0} \left(\sup_{x \in E} \frac{\mu(B(x, br))}{\mu(B(x, r))} \right)$$

$$\leq \limsup_{r \downarrow 0} \left(\sup_{x \in E} \frac{\mu(B(x, a^n r))}{\mu(B(x, r))} \right)$$

$$= \limsup_{r \downarrow 0} \left(\sup_{x \in E} \prod_{i=1}^{n} \frac{\mu(B(x, a^i r))}{\mu(B(x, a^{i-1} r))} \right)$$

$$\leq \left(T_a(E) \right)^n < \infty.$$

$\square$

In the sequel we denote

$$P_0(\mathbb{R}^d, E) = \{ \mu \in P(\mathbb{R}^d), \text{ such that, } \exists a > 1; \ \forall x \in E, \ T_a(x) < \infty \},$$

$$P_1(\mathbb{R}^d, E) = \{ \mu \in P(\mathbb{R}^d), \text{ such that, } \exists a > 1; \ T_a(E) < \infty \},$$

$$P_0(\mathbb{R}^d) = P_0(\mathbb{R}^d, Support(\mu)),$$

and

$$P_1(\mathbb{R}^d) = P_1(\mathbb{R}^d, Support(\mu)).$$

For $\delta > 0$, and $q \in \mathbb{R}$, we write

$$\mathcal{S}^q_{\mu, \delta}(E) = \sup \left\{ \sum_i (\mu(B(x_i, \delta)))^q; \ (B(x_i, \delta))_i \text{ packing of } E \right\},$$

$$\mathcal{T}_{\mu,\delta}^q(E) = \inf\left\{\sum_i \left(\mu(B(x_i,\delta))\right)^q; \; \left(B(x_i,\delta)\right)_i \text{ covering of } E\right\},$$

$$\overline{C}_\mu^q(E) = \limsup_{\delta\downarrow 0} \frac{\log\left(\mathcal{S}_{\mu,\delta}^q(E)\right)}{-\log\delta},$$

$$\underline{C}_\mu^q(E) = \liminf_{\delta\downarrow 0} \frac{\log\left(\mathcal{S}_{\mu,\delta}^q(E)\right)}{-\log\delta},$$

$$\overline{L}_\mu^q(E) = \limsup_{\delta\downarrow 0} \frac{\log\left(\mathcal{T}_{\mu,\delta}^q(E)\right)}{-\log\delta},$$

and

$$\underline{L}_\mu^q(E) = \liminf_{\delta\downarrow 0} \frac{\log\left(\mathcal{T}_{\mu,\delta}^q(E)\right)}{-\log\delta}.$$

We have the following result.

Theorem 8.1.

(1) *For all $q \leq 0$, we have*
   (i) *$b_{\mu,E}(q) \leq \underline{L}_\mu^q(E) = \underline{C}_\mu^q(E)$.*
   (ii) *$\overline{L}_{\mu,E}(q) = \overline{C}_\mu^q(E) = \Lambda_{\mu,E}(q)$.*

(2) *For all $q > 0$, we have*
   (i) *$\underline{L}_\mu^q(E) \leq \underline{C}_\mu^q(E)$.*
   (ii) *$\overline{L}_{\mu,E}(q) \leq \overline{C}_\mu^q(E) \leq \Lambda_{\mu,E}(q)$.*

(3) *If $\mu \in P_0(\mathbb{R}^d)$, and $q > 0$, we have*

$$b_{\mu,E}(q) \leq \overline{L}_\mu^q(E).$$

(4) *If $\mu \in P_1(\mathbb{R}^d)$, and $q > 0$, we have*
   (i) *$\underline{L}_\mu^q(E) = \underline{C}_\mu^q(E)$.*
   (ii) *$\overline{L}_{\mu,E}(q) = \overline{C}_\mu^q(E) = \Lambda_{\mu,E}(q)$.*

Proof. **(1.i)** We will prove, in the first step, that for all $q \leq 0$,

$$\underline{L}_\mu^q(E) = \underline{C}_\mu^q(E), \quad \text{and} \quad \overline{L}_\mu^q(E) = \overline{C}_\mu^q(E).$$

According to Besicovitch's covering Theorem, there exists $A > 0$, such that, for all $q \in \mathbb{R}$,

$$\mathcal{T}_{\mu,\delta}^q(E) \leq A\mathcal{S}_{\mu,\delta}^q(E).$$

Hence, $\forall\, q \in \mathbb{R}$, we get

$$\underline{L}^q_\mu(E) \leq \underline{C}^q_\mu(E), \quad \text{and} \quad \overline{L}^q_\mu(E) \leq \overline{C}^q_\mu(E).$$

This proves assertion (2.i) and the first inequality of assertion (2.ii). Conversely, consider a centred δ-packing $\big(B(x_i, \delta)\big)_i$ of E. Let $\big(B(y_i, \frac{\delta}{2})\big)$ be a centred $\frac{\delta}{2}$-covering E. For all $i \in \mathbb{N}$, let k_i be, such that, $x_i \in B(y_{k_i}, \frac{\delta}{2})$. It holds that $\big(B(x_i, \delta)\big)_i$ is a centred δ-packing of E. Moreover, for $i \neq j$, we have $k_i \neq k_j$. So, as $q \leq 0$, we get

$$\sum_i \big(\mu(B(x_i, \delta))\big)^q = \sum_i \left(\frac{\mu(B(x_i, \delta))}{\mu(B(y_{k_i}, \delta/2))} \right)^q \big(\mu(B(y_{k_i}, \tfrac{\delta}{2}))\big)^q$$

$$\leq \sum_i \big(\mu(B(y_i, \tfrac{\delta}{2}))\big)^q.$$

Consequently,

$$\mathcal{S}^q_{\mu,\delta}(E) \leq \mathcal{T}^q_{\mu, \frac{\delta}{2}}(E).$$

Thus, $\forall\, q \leq 0$, it holds that

$$\underline{L}^q_\mu(E) \geq \underline{C}^q_\mu(E), \quad \text{and} \quad \overline{L}^q_\mu(E) \geq \overline{C}^q_\mu(E).$$

Now, assume that the right hand quantities in assertion (**1.i**) are finite. (The inequality is evident if these members are infinite). Let $t > \underline{L}^q_\mu(E)$, and $F \subseteq E$. There exists a sequence $(\delta_n)_n \subseteq]0,1[$, decreasing to 0, such that,

$$t > \frac{\log(\mathcal{T}^q_{\mu,\delta_n}(E))}{-\log \delta_n}, \quad \forall\, n \in \mathbb{N}.$$

As a result, $\forall\, n \in \mathbb{N}$, there exist a centred δ_n-covering $\big(B(x_{ni}, \delta_n)\big)_i$ of E, such that,

$$\sum_i \big(\mu(B(x_{ni}, \delta_n))\big)^q < \delta_n^{-t}.$$

Without loss of generality, we may assume that all the balls meet the set F. For each i fixed, choose an element $y_i \in B(x_{ni}, \delta_n) \cap F$. Hence, $\big(B(y_i, 2\delta_n)\big)_i$ is a centred δ_n-covering of F. Therefore,

$$\overline{\mathcal{H}}^{q,t}_{\mu,2\delta_n}(F) \leq \sum_i \big(\mu(B(x_{ni}, \delta_n))\big)^q (4\delta_n)^t$$

$$= 4^t \sum_i \left(\frac{\mu(B(y_i, 2\delta_n))}{\mu(B(x_{ni}, \delta_n))} \right)^q \big(\mu(B(x_{ni}, \delta_n))\big)^q \delta_n^t$$

$$\leq 4^t \sum_i \big(\mu(B(x_{ni}, \delta_n))\big)^q \delta_n^t$$

$$\leq 4^t \delta_n^{-t} \delta_n^t = 4^t,$$

which gives

$$\overline{\mathcal{H}}_{\mu}^{q,t}(F) \leq 4^t, \quad \forall F \subseteq E, \text{ and } t > \underline{L}_{\mu}^q(E).$$

Consequently,

$$\mathcal{H}_{\mu}^{q,t}(E) \leq 4^t < \infty, \quad \forall t > \underline{L}_{\mu}^q(E).$$

Finally, we get

$$b_{\mu,E}(q) \leq t, \quad \forall t > \underline{L}_{\mu}^q(E),$$

which leads to

$$b_{\mu,E}(q) \leq \underline{L}_{\mu}^q(E).$$

(1.ii) It remains to prove the last equality. Let's show that

$$\overline{C}_{\mu}^q(E) \leq \Lambda_{\mu,E}(q).$$

The inequality is satisfied if the right member is infinite. Suppose, so, that $\Lambda_{\mu,E}(q) < +\infty$. Let $t = \Lambda_{\mu,E}(q)$, $\epsilon > 0$, and $0 < \delta_\epsilon < 1$ be, such that,

$$\overline{\mathcal{P}}_{\mu,\delta}^{q,t+\epsilon}(E) < 1, \forall 0 < \delta < \delta_\epsilon.$$

For a centred δ-packing $\big(B(x_i,\delta)\big)_i$ of E. we get

$$\sum_i \big(\mu(B(x_i,\delta))\big)^q = (2\delta)^{-(t+\epsilon)} \sum_i \big(\mu(B(x_i,\delta))\big)^q (2\delta)^{t+\epsilon}$$
$$\leq (2\delta)^{-(t+\epsilon)} \overline{\mathcal{P}}_{\mu,\delta}^{q,t+\epsilon}(E)$$
$$\leq (2\delta)^{-(t+\epsilon)}.$$

Consequently,

$$\mathcal{S}_{\mu,\delta}^q(E) \leq (2\delta)^{-(t+\epsilon)},$$

which gives that

$$\overline{C}_{\mu}^q(E) \leq \Lambda_{\mu,E}(q), \forall q \in \mathbb{R}.$$

Remark that this gives also the last inequality in assertion **(2.ii)**. We shall next show that

$$\Lambda_{\mu,E}(q) \leq \overline{C}_{\mu}^q(E).$$

Suppose, as usually, that the right member is finite. Let $t = \Lambda_{\mu,E}(q)$, $\epsilon > 0$, and $0 < \delta_0 < 1$. From the definition of $\Lambda_{\mu,E}(q)$, we have

$$\infty = \overline{\mathcal{P}}_{\mu}^{q,t+\epsilon/2}(E) \leq \overline{\mathcal{P}}_{\mu,\delta_0}^{q,t-\epsilon/2}(E).$$

There exists a centred δ_0-packing $\big(B(x_i, r_i)\big)_i$ of E, such that,

$$1 < \sum_i \big(\mu(B(x_i, \delta))\big)^q (2r_i)^{t-\epsilon/2}.$$

For $n \in \mathbb{N}$, Let

$$I_n = \{\, i \in \mathbb{N};\ \frac{\delta}{2^{n+1}} \leq r_i < \frac{\delta}{2^n}\,\}, \quad \text{and} \quad \mu_n = \sum_{i \in I_n} \big(\mu(B(x_i, \delta))\big)^q.$$

We have

$$1 < \sum_i \big(\mu(B(x_i, \delta))\big)^q (2r_i)^{t-\epsilon/2}$$

$$\leq 2^{t-\epsilon/2}\Big(1 \vee (\tfrac{1}{2})^{t-\epsilon/2}\Big) \sum_n \mu_n (\frac{\delta_0}{2^n})^{t-\epsilon/2}$$

$$\leq 2^{t-\epsilon/2}\Big(1 \vee (\tfrac{1}{2})^{t-\epsilon/2}\Big) \sum_n \mu_n (\frac{\delta_0}{2^n})^{t-\epsilon}(\frac{\delta_0}{2^n})^{\epsilon/2}$$

$$\leq 2^{t-\epsilon/2}\Big(1 \vee (\tfrac{1}{2})^{t-\epsilon/2}\Big) \sup_m \Big(\mu_m (\frac{\delta_0}{2^m})^{t-\epsilon}\Big) \sum_n (\frac{\delta_0}{2^n})^{\epsilon/2}$$

$$\leq C \sup_m \Big(\mu_m (\frac{\delta_0}{2^m})^{t-\epsilon}\Big),$$

where

$$C = 2^{t-\epsilon/2}\Big(1 \vee (\tfrac{1}{2})^{t-\epsilon/2}\Big) \sum_n (\frac{\delta_0}{2^n})^{\epsilon/2}.$$

Let next $N \in \mathbb{N}$ be, such that,

$$1 < C\mu_N(\frac{\delta_0}{2^N})^{t-\epsilon}, \quad \text{and} \quad \delta = \frac{\epsilon_0}{2^{N+1}}.$$

It follows that $\big(B(x_i, \delta)\big)_i$ is a centred δ-packing of E. Consequently,

$$S^q_{\mu,\delta}(E) \geq \sum_{i \in I_N} \big(\mu(B(x_i, \delta))\big)^q$$

$$\geq \sum_{i \in I_N} \left(\frac{\mu(B(x_i, \frac{\delta_0}{2^{N+1}}))}{\mu(B(x_i, \frac{\delta_0}{2^N}))}\right)^q \big(\mu(B(x_i, r_i))\big)^q$$

$$\geq \sum_{i \in I_N} \big(\mu(B(x_i, r_i))\big)^q$$

$$\geq \mu_N > C^{-1}(\frac{\delta_0}{2^N})^{-(t-\epsilon)}.$$

Hence,

$$\overline{C}_\mu^q(E) \geq \Lambda_{\mu,E}(q), \ \forall q \leq 0.$$

(2) is already shown into the proof of assertion 1.

(3) Let $t = \overline{L}_\mu^q(E)$, and $F \subseteq E_m$, where

$$E_m = \left\{ x \in E; \ \frac{\mu(B(x_i, 3r))}{\mu(B(x_i, r))} < m, \ 0 < r < \frac{1}{m} \right\}.$$

There exists a sequence $(\delta_n)_n \in]0, 1[$ decreasing to 0, such that,

$$t < \frac{\log(\mathcal{T}_{\mu,\delta_n}^q(F))}{-\log \delta_n}, \quad \forall n \in \mathbb{N}.$$

This yields that

$$\overline{\mathcal{H}}_{\mu,\delta_n}^{q,t}(F) \leq 4^t m^q, \quad \forall n, m \ \text{and} \ \forall F \subseteq E_m,$$

which implies that

$$\overline{\mathcal{H}}_\mu^{q,t}(F) \leq 4^t m^q, \quad \forall m \ \text{and} \ \forall F \subseteq E_m.$$

By taking the sup on $F \subseteq E_m$, we obtain

$$\mathcal{H}_\mu^{q,t}(E_m) \leq 4^t m^q < \infty, \quad \forall m \ \text{and} \ \forall t > \overline{L}_\mu^q(E).$$

Hence,

$$dim\,(E_m) \leq t, \quad \forall m, \ \text{and} \ \forall\, t > \overline{L}_\mu^q(E).$$

As the sequence $(E_m)_m$ increases to E, we obtain

$$b_{\mu,E}(q) \leq t, \quad \forall\, t > \overline{L}_\mu^q(E),$$

which yields that

$$b_{\mu,E}(q) \leq \overline{L}_\mu^q(E).$$

(4.i) It remains to show that

$$\underline{L}_\mu^q(E) \geq \underline{C}_\mu^q(E) \quad \text{and} \quad \overline{L}_\mu^q(E) \geq \overline{C}_\mu^q(E) \geq \Lambda_{\mu,E}(q).$$

So, recall that $\mu \in P_1(\mathbb{R}^d, E)$. Consequently, there exists $\beta > 0$, and $r_0 > 0$, such that,

$$\frac{\mu(B(x, 3r))}{\mu(B(x, r))} \leq \beta, \ \forall\, 0 < r < r_0 \ \text{and} \ \forall\, x \in E.$$

Let next $0 < \delta < r_0$, $(B(x_i, \delta))_i$ be a δ-packing of E, and $(B(y_i, \frac{\delta}{2}))_i$ a centred $\frac{\delta}{2}$-covering of E. For $i \in \mathbb{N}$, let $k_i \in \mathbb{N}$ be, such that, $x_i \in B(y_{k_i}, \frac{\delta}{2})$. It follows that

$$\sum_i \mu(B(x_i, \delta)) = \sum_i \left(\frac{\mu(B(x_i, \delta))}{\mu(B(y_{k_i}, \frac{\delta}{2}))} \right)^q \left(\mu(B(y_{k_i}, \frac{\delta}{2})) \right)^q$$

$$\leq \sum_i \left(\frac{\mu(B(y_{k_i}, \frac{3\delta}{2}))}{\mu(B(y_{k_i}, \frac{\delta}{2}))} \right)^q \left(\mu(B(y_{k_i}, \frac{\delta}{2})) \right)^q$$

$$\leq \beta^q \sum_i \left(\mu(B(y_{k_i}, \frac{\delta}{2})) \right)^q,$$

which leads to

$$S_{\mu, \delta}^q(E) \leq \beta^q T_{\mu, \frac{\delta}{2}}^q(E).$$

Consequently,

$$\underline{L}_\mu^q(E) \geq \underline{C}_\mu^q(E) \quad \text{and} \quad \overline{L}_\mu^q(E) \geq \overline{C}_\mu^q(E).$$

We shall now prove that

$$\overline{C}_\mu^q(E) \geq \Lambda_{\mu, E}(q).$$

Indeed, μ being in $P_1(\mathbb{R}^d, E)$, there exists consequently $\beta > 0$, and $0 < r_0 < 1$, satisfying

$$\frac{\mu(B(x, 2r))}{\mu(B(x, r))} \leq \beta; \quad \forall \, 0 < r < r_0 \text{ and } \forall x \in E.$$

Let nest $t = \Lambda_{\mu, E}(q)$, $\epsilon > 0$, and $0 < \delta_0 < r_0$. It follows that

$$\overline{\mathcal{P}}_\mu^{q, t - \epsilon/2}(E) = \infty.$$

As a result, there exists a δ_0-packing $(B(x_i, r_i))_i$ of E, for which, we have

$$1 < \sum_i \left(\mu(B(x_i, r_i)) \right)^q (2r_i)^{t - \epsilon/2}.$$

Consider next I_N, μ_N, and δ defined as in the proof of assertion (**1.ii**). We have

$$S_{\mu, \delta}^q(E) \geq \sum_{i \in I_N} \left(\mu(B(x_i, \delta)) \right)^q$$

$$\geq \beta^{-q} \sum_{i \in I_N} \left(\frac{\mu(B(x_i, \delta_0/2^{N+1}))}{\mu(B(x_i, \delta_0/2^N))} \right)^q \left(\mu(B(x_i, r_i)) \right)^q$$

$$\geq \beta^{-q} \sum_{i \in I_N} \left(\mu(B(x_i, r_i)) \right)^q$$

$$\geq \beta^{-q} \mu_N > \beta^{-q} C^{-1} (\frac{\delta_0}{2^N})^{-(t - \epsilon)}.$$

Consequently,

$$\overline{C}^q_\mu(E) \geq \Lambda_{\mu,E}(q).$$

$\square$

In the sequel, for $q \in \mathbb{R}*$, and $_d > 0$, we denote

$$I^q_{\mu,\delta} = \frac{1}{q} \log \int_{Support(\mu)} \big(\mu(B(x,\delta))\big)^q d\mu(x),$$

$$\overline{I}^q_\mu = \limsup_{\delta \downarrow 0} \frac{I^q_{\mu,\delta}}{-\log \delta},$$

and

$$\underline{I}^q_\mu = \liminf_{\delta \downarrow 0} \frac{I^q_{\mu,\delta}}{-\log \delta}.$$

WE have the following result.

Proposition 8.6. *The following assertions hold.*

(1) $\forall\, q \in \mathbb{R}^*$, $\mu \in P_1(\mathbb{R}^d)$ *we have*
     (i) $\underline{C}^{q+1}_\mu(Support(\mu)) = q\underline{I}^q_\mu$.
     (ii) $\overline{C}^{q+1}_\mu(Support(\mu)) = q\overline{I}^q_\mu$.

(2) $\forall\, q < 0$ *we have*
     (i) $q\underline{I}^q_\mu \leq \underline{L}^{q+1}_\mu(Support(\mu))$.
     (ii) $q\overline{I}^q_\mu \leq \overline{L}^{q+1}_\mu(Support(\mu))$.

(3) $\forall\, q > 0$ *we have*
     (i) $\underline{C}^{q+1}_\mu(Support(\mu)) \leq q\underline{I}^q_\mu$.
     (ii) $\overline{C}^{q+1}_\mu(Support(\mu)) \leq q\overline{I}^q_\mu$.

Proof. (1) Assume that $q < 0$. As $\mu \in P_1(\mathbb{R}^d)$, there exists $A > 0$, and $r_0 > 0$, for which we have

$$\frac{\mu(B(x,2r))}{\mu(B(x,r))} < A\,; \quad \forall\, x \in Support(\mu), \quad \text{and} \quad 0 < r < r_0.$$

Let $0 < \delta < r_0$, and $(B(x_i,\delta))_i$ be a δ-packing of $Support(\mu)$. We have

$$\sum_i \big(\mu(B(x_i,\delta))\big)^{q+1} = \sum_i \big(\mu(B(x_i,\delta))\big)^q \int_{B(x_i,\delta)} d\mu(x)$$

$$\leq A^{-q} \sum_i \int_{B(x_i,\delta)} \big(\mu(B(x,2\delta))\big)^q d\mu(x)$$

$$\leq A^{-q} \int_{Support(\mu)} \big(\mu(B(x,2\delta))\big)^q d\mu(x).$$

Therefore,

$$S_{\mu,\delta}^{q+1}(Support(\mu)) \leq A^{-q} \exp(qI_{\mu,2\delta}^q).$$

As a result,

$$\underline{C}_\mu^{q+1}(Support(\mu)) \leq q\underline{I}_\mu^q \quad \text{and} \quad \overline{C}_\mu^{q+1}(Support(\mu)) \leq q\overline{I}_\mu^q.$$

Conversely, for $\delta > 0$, let $\left(B(x_i,\delta)\right)_i$ be a centred δ-covering of $Support(\mu)$. Let also $\left(B(x_{ij},\delta)\right)_{1\leq i\leq \xi,j}$ be the ξ families defined in the Besicovitch's covering Theorem. We have

$$\sum_{i,j}\left(\mu(B(x_{ij},\delta))\right)^{q+1} = \sum_{i,j}\left(\mu(B(x_{ij},\delta))\right)^q \int_{B(x_{ij},\delta)} d\mu(x)$$

$$\geq \sum_{i,j}\int_{B(x_{ij},\delta)}\left(\mu(B(x,2\delta))\right)^q d\mu(x)$$

$$\geq \int_{Support(\mu)}\left(\mu(B(x,2\delta))\right)^q d\mu(x).$$

Consequently,

$$\xi S_{\mu,\delta}^{q+1}(Support(\mu)) \geq \exp(qI_{\mu,2\delta}^q).$$

As a result, we obtain

$$\underline{C}_\mu^{q+1}(Support(\mu)) \geq q\underline{I}_\mu^q \quad \text{and} \quad \overline{C}_\mu^{q+1}(Support(\mu)) \geq q\overline{I}_\mu^q.$$

Assume now that $q > 0$. Let $\delta > 0$, and $\left(B(x_i,\delta)\right)_i$ be a δ-packing of $Support(\mu)$. We have

$$\sum_i\left(\mu(B(x_i,\delta))\right)^{q+1} = \sum_i\left(\mu(B(x_i,\delta))\right)^q \int_{B(x_i,\delta)} d\mu(x)$$

$$\leq \sum_i\int_{B(x_i,\delta)}\left(\mu(B(x,2\delta))\right)^q d\mu(x)$$

$$\leq \int_{Support(\mu)}\left(\mu(B(x,2\delta))\right)^q d\mu(x).$$

It results that

$$S_{\mu,\delta}^{q+1}(Support(\mu)) \leq \exp(qI_{\mu,2\delta}^q).$$

As a consequence,

$$\underline{C}_\mu^{q+1}(Support(\mu)) \leq q\underline{I}_\mu^q \quad \text{and} \quad \overline{C}_\mu^{q+1}(Support(\mu)) \leq q\overline{I}_\mu^q.$$

Conversely, as $\mu \in P_1(\mathbb{R}^d)$, there exists $A > 0$, and $r_0 > 0$, satisfying

$$\frac{\mu(B(x,2r))}{\mu(B(x,r))} < A \; ; \forall \; x \in Support(\mu) \text{ and } 0 < r < r_0.$$

Let $0 < \delta < r_0$, $(B(x_i, \delta))_i$ be a δ-covering of $Support(\mu)$, and $\left(B(x_{ij}, \delta)\right)_{1 \leq i \leq \xi, j}$ the ξ families defined in the Besicovitch's covering Theorem. We get

$$\sum_{i,j} \left(\mu(B(x_{ij}, \delta))\right)^{q+1} = \sum_{i,j} \left(\mu(B(x_{ij}, \delta))\right)^q \int_{B(x_{ij}, \delta)} d\mu(x)$$

$$\geq A^{-q} \sum_{i,j} \int_{B(x_{ij}, \delta)} \left(\mu(B(x, 2\delta))\right)^q d\mu(x)$$

$$\geq A^{-q} \int_{Support(\mu)} \left(\mu(B(x, 2\delta))\right)^q d\mu(x).$$

As a consequence,

$$\xi \mathcal{S}_{\mu,\delta}^{q+1}(Support(\mu)) \geq A^{-q} \exp(q I_{\mu,2\delta}^q).$$

Therefore,

$$\underline{C}_\mu^{q+1}(Support(\mu)) \geq q \underline{I}_\mu^q \quad \text{and} \quad \overline{C}_\mu^{q+1}(Support(\mu)) \geq q \overline{I}_\mu^q.$$

Remark that we have also proved assertion **3)**.

(2) Let $q < 0$, $\delta > 0$, and $\left(B(x_i, \delta)\right)_i$ be a δ-covering of $Support(\mu)$. We have

$$\sum_i \left(\mu(B(x_i, \delta))\right)^{q+1} = \sum_i \left(\mu(B(x_i, \delta))\right)^q \int_{B(x_i, \delta)} d\mu(x)$$

$$\geq \sum_i \int_{B(x_i, \delta)} \left(\mu(B(x, 2\delta))\right)^q d\mu(x)$$

$$\geq \int_{Support(\mu)} \left(\mu(B(x, 2\delta))\right)^q d\mu(x).$$

It follows that

$$\mathcal{T}_{\mu,\delta}^{q+1}(Support(\mu)) \geq \exp(q I_{\mu,2\delta}^q).$$

As a result,

$$\underline{L}_\mu^{q+1}(Support(\mu)) \geq q \underline{I}_\mu^q \quad \text{and} \quad \overline{L}_\mu^{q+1}(Support(\mu)) \geq q \overline{I}_\mu^q.$$

$\square$

8.3 The multifractal spectrum

This section is concerned with the computation of the spectrum of singularities for measures, in the framework of the generalized multifractal analysis. In this context, the possibility of being Gibbs measures may fail, compared to Chapter 7.

Let μ be a Borel probability measure on $\mathbb{R}^d$, $x \in \mathbb{R}^d$, and put

$$\underline{\alpha}_\mu(x) = \liminf_{r \downarrow 0} \frac{\log(\mu(B(x,r)))}{\log r},$$

and

$$\overline{\alpha}_\mu(x) = \limsup_{r \downarrow 0} \frac{\log(\mu(B(x,r)))}{\log r}.$$

These quantities define, respectively, the local lower dimension, and the local upper dimension of μ at the point x. In the case of equality, the common value is called the local dimension of μ in x, and will be denoted $\alpha_\mu(x)$. In the sequel, we consider, for all $\alpha \in \mathbb{R}_+$, the sets

$$\underline{X}_\alpha = \{x \in Support(\mu); \ \underline{\alpha}_\mu(x) \geq \alpha\},$$

called the lower α-singularity set of the measure μ. Consider also

$$\overline{X}^\alpha = \{x \in Support(\mu); \ \overline{\alpha}_\mu(x) \leq \alpha\},$$

called, similarly, the upper α-singularity set of the measure μ. Their intersection

$$X(\alpha) = \underline{X}_\alpha \cap \overline{X}^\alpha,$$

defines finally the α-singularity set of the measure μ. We will show that these sets are in the heart of the computation of the spectrum of singularities of μ. Indeed, we have the following definition.

Definition 8.2. We call multifractal of a measure μ the function

$$d: \alpha \mapsto d(\alpha) = dim\, X(\alpha),$$

where dim is the Hausdorff dimension.

In this chapter, we will study the link between this spectrum, and the functions b_μ, B_μ, and Λ_μ introduced previously.

Proposition 8.7. *Consider $\alpha \geq 0$, and $q \in \mathbb{R}$. The following assertions hold.*

(1) If $\alpha q + b_\mu(q) \geq 0$, we have

    (i) $dim\overline{X}^\alpha \leq \alpha q + b_\mu(q)$ for $q \geq 0$

    (ii) $dim\underline{X}_\alpha \leq \alpha q + b_\mu(q)$ for $q \leq 0$

(2) If $\alpha q + B_\mu(q) \geq 0$, we have

    (i) $Dim\overline{X}^\alpha \leq \alpha q + B_\mu(q)$ for $q \geq 0$

    (ii) $Dim\underline{X}_\alpha \leq \alpha q + B_\mu(q)$ for $q \leq 0$

Proof. (1.i) It suffices to prove that

$$(*) \begin{cases} \forall\, \delta > 0,\ t \in \mathbb{R},\ \text{and}\ q \geq 0,\ \alpha q + t \geq 0,\ \text{we have} \\ \mathcal{H}^{\alpha q + t + \delta}(\overline{X}^\alpha) \leq 2^{\alpha q + \delta} \mathcal{H}_\mu^{q,t}(\overline{X}^\alpha). \end{cases}$$

Indeed, if $(*)$ is true, we get

$$\mathcal{H}^{\alpha q + t + \delta}(\overline{X}^\alpha) = 0, \quad \forall\, t > b_\mu(q),\ \delta > 0.$$

Hence,

$$dim\overline{X}^\alpha \leq \alpha q + t + \delta, \quad \forall\, t > b_\mu(q),\ \delta > 0,$$

which leads to

$$dim\overline{X}^\alpha \leq \alpha q + b_\mu(q).$$

We will now prove the property $(*)$. It is clear that this property is true for all $q = 0$. Assume so, that $q < 0$. For $m \in \mathbb{N}^*$, consider the set

$$\overline{X}_m^\alpha = \{\, x \in \overline{X}^\alpha;\ \frac{\log(\mu(B(x,r)))}{\log r} \leq \alpha + \frac{\delta}{q};\ 0 < r < \frac{1}{m} \,\}.$$

Let $0 < \eta < \dfrac{1}{m}$, and $(B(x_i, r_i))_i$ be a η-covering of $\overline{X}_m^\alpha$. As

$$\frac{\log(\mu(B(x_i, r_i)))}{\log r_i} \leq \alpha + \frac{\delta}{q},$$

we have

$$(\mu(B(x_i, r_i)))^q (2r_i)^t \geq 2^t r_i^{\alpha q + t + \delta}.$$

Hence,

$$\mathcal{H}_\eta^{\alpha q + t + \delta}(\overline{X}_m^\alpha) \leq \sum_i (2r_i)^{\alpha q + t + \delta} \leq 2^{\alpha q + \delta} \sum_i (\mu(B(x_i, r_i)))^q (2r_i)^t,$$

which yields that

$$\mathcal{H}_\eta^{\alpha q + t + \delta}(\overline{X}_m^\alpha) \leq 2^{\alpha q + \delta} \overline{\mathcal{H}}_{\mu,\eta}^{q,t}(\overline{X}_m^\alpha), \quad \forall \eta > 0.$$

Consequently,

$$\mathcal{H}^{\alpha q+t+\delta}(\overline{X}^\alpha_m) \leq 2^{\alpha q+\delta}\overline{\mathcal{H}}^{q,t}_\mu(\overline{X}^\alpha_m) \leq 2^{\alpha q+\delta}\mathcal{H}^{q,t}_\mu(\overline{X}^\alpha_m).$$

As a result,

$$\mathcal{H}^{\alpha q+t+\delta}(\overline{X}^\alpha_m) \leq 2^{\alpha q+\delta}\mathcal{H}^{q,t}_\mu(\overline{X}^\alpha_m).$$

Finally, the limit on m gives

$$\mathcal{H}^{\alpha q+t+\delta}(\overline{X}^\alpha) \leq 2^{\alpha q+\delta}\mathcal{H}^{q,t}_\mu(\overline{X}^\alpha).$$

(1.ii) An analogue process to assertion **(1.i)** allows to establish the following property

$$\begin{cases} \forall\, \delta > 0,\ t \in \mathbb{R},\ \text{and } q \leq 0,\ \alpha q + t \geq 0,\ \text{we have} \\ \mathcal{H}^{\alpha q+t+\delta}(\underline{X}_\alpha) \leq 2^{\alpha q+\delta}\mathcal{H}^{q,t}_\mu(\underline{X}_\alpha), \end{cases}$$

which in turn leads to

$$dim\underline{X}_\alpha \leq \alpha q + t + \delta, \quad \forall\, t > b_\mu(q),\ \delta > 0.$$

So finally,

$$dim\underline{X}_\alpha \leq \alpha q + b_\mu(q).$$

(2.i) Let's start by establishing the property

$$(**) \begin{cases} \forall\, \delta > 0,\ t \in \mathbb{R},\ \text{and } q \geq 0,\ \alpha q + t \geq 0,\ \text{we have} \\ \mathcal{P}^{\alpha q+t+\delta}(\overline{X}^\alpha) \leq 2^{\alpha q+\delta}\mathcal{P}^{q,t}_\mu(\overline{X}^\alpha). \end{cases}$$

Such property is obvious for $q = 0$. Suppose then that $q > 0$. For $m \in \mathbb{N}$, Consider the set $\overline{X}^\alpha_m$ defined in assertion **1)**, and Let $E \subseteq \overline{X}^\alpha_m, 0 < \eta < \dfrac{1}{m}$, and $(B(x_i, r_i))_i$ be a η-packing of E. We have

$$(\mu(B(x_i, r_i)))^q (2r_i)^t \geq 2^t r_i^{\alpha q+t+\delta}.$$

Therefore,

$$\sum_i (2r_i)^{\alpha q+t+\delta} \leq 2^{\alpha q+\delta}\sum_i (\mu(B(x_i, r_i)))^q (2r_i)^t \leq 2^{\alpha q+\delta}\overline{\mathcal{P}}^{q,t}_{\mu,\eta}(E).$$

Consequently,

$$\overline{\mathcal{P}}^{\alpha q+t+\delta}_\eta(E) \leq 2^{\alpha q+\delta}\overline{\mathcal{P}}^{q,t}_{\mu,\eta}(E), \quad \forall\, \eta > 0,$$

which leads to

$$\overline{\mathcal{P}}^{\alpha q+t+\delta}(E) \leq 2^{\alpha q+\delta}\overline{\mathcal{P}}^{q,t}_\mu(E), \quad \forall\, E \subseteq \overline{X}^\alpha_m.$$

Therefore, if $(E_i)_i$ is any covering of $\overline{X}_m^\alpha$, we obtain

$$\mathcal{P}^{\alpha q+t+\delta}(\overline{X}_m^\alpha) = \mathcal{P}^{\alpha q+t+\delta}\left(\bigcup_i (\overline{X}_m^\alpha \cap E_i)\right)$$

$$= \sum_i \mathcal{P}^{\alpha q+t+\delta}\left(\overline{X}_m^\alpha \cap E_i\right)$$

$$\leq \sum_i \overline{\mathcal{P}}^{\alpha q+t+\delta}\left(\overline{X}_m^\alpha \cap E_i\right)$$

$$\leq 2^{\alpha q+\delta} \sum_i \overline{\mathcal{P}}_\mu^{q,t}\left(\overline{X}_m^\alpha \cap E_i\right)$$

$$\leq 2^{\alpha q+\delta} \sum_i \overline{\mathcal{P}}_\mu^{q,t}(E_i).$$

We deduce that

$$\mathcal{P}^{\alpha q+t+\delta}(\overline{X}_m^\alpha) \leq 2^{\alpha q+\delta}\mathcal{P}_\mu^{q,t}(\overline{X}_m^\alpha).$$

Consequently,

$$\mathcal{P}^{\alpha q+t+\delta}(\overline{X}^\alpha) \leq 2^{\alpha q+\delta}\mathcal{P}_\mu^{q,t}(\overline{X}^\alpha).$$

Therefore,

$$\mathcal{P}^{\alpha q+t+\delta}(\overline{X}^\alpha), \quad \forall\, t > B_\mu,\ \delta > 0.$$

It results that

$$Dim\overline{X}^\alpha \leq \alpha q + t + \delta, \quad \forall\, t > B_\mu,\ \delta > 0,$$

and finally,

$$Dim\overline{X}^\alpha \leq \alpha q + B_\mu.$$

(2.ii) By an analogue process to the previous assertion, we get

$$\begin{cases} \forall\, \delta > 0,\ t \in \mathbb{R},\ \text{and}\ q \leq 0,\ \alpha q + t \geq 0\ \text{we have} \\ \mathcal{P}^{\alpha q+t+\delta}(\underline{X}_\alpha) \leq 2^{\alpha q+\delta}\mathcal{P}_\mu^{q,t}(\underline{X}_\alpha), \end{cases}$$

which gives that

$$\mathcal{P}^{\alpha q+t+\delta}(\underline{X}_\alpha) = 0, \quad \forall\, t > B_\mu(q),\ \delta > 0.$$

As a result, we get

$$Dim\underline{X}_\alpha \leq \alpha q + t + \delta, \quad \forall\, t > B_\mu(q),\ \delta > 0,$$

and finally,

$$Dim\underline{X}_\alpha \leq \alpha q + B_\mu(q).$$

$\square$

Proposition 8.8. $\forall \, q \in \mathbb{R}$, *such that,* $\alpha q + b_\mu(q) < 0$ *or* $\alpha q + B_\mu(q) < 0$, *we have* $X(\alpha) = \emptyset$.

Proof. It suffices to prove that whenever $\alpha q + b_\mu(q) < 0$ or $\alpha q + B_\mu(q) < 0$, we get $\underline{X}_\alpha = \emptyset$, for $q \leq 0$, and $\overline{X}^\alpha = \emptyset$ for $q \geq 0$. Indeed, let $q \leq 0$, and suppose that $\underline{X}_\alpha \neq \emptyset$. There exists $x \in Support(\mu)$, satisfying $\underline{\alpha}_\mu(x) \geq \alpha$. Then, for all $\epsilon > 0$, there exists a sequence $(r_n)_n$ decreasing to 0, such that,

$$0 < r_n < \frac{1}{n}, \quad \text{and} \quad \mu(B(x, r_n)) < r_n^{\alpha - \epsilon}.$$

Consequently,

$$\Big(\mu(B(x, r_n))\Big)^q (2r_n)^t > 2^t, \quad \text{where } t = -q(\alpha - \epsilon),$$

which leads to $\mathcal{H}_\mu^{q,t}(\{x\}) > 2^t$. As a result,

$$b_\mu(q) \geq dim_\mu^q(\{x\}) \geq t, \quad \forall \, \epsilon > 0.$$

Letting $\epsilon \to 0$, we obtain $b_\mu(q) \geq -\alpha t$ which is contradictory.
Let now $q \geq 0$, and suppose that $\overline{X}^\alpha \neq \emptyset$. There exists $x \in Support(\mu)$, such that, $\overline{\alpha}_\mu(x) \leq \alpha$. Then, for all $\epsilon > 0$, there exists a sequence $(r_n)_n$ decreasing to 0, such that,

$$0 < r_n < \frac{1}{n} \quad \text{and} \quad \mu(B(x, r_n)) > r_n^{\alpha + \epsilon}.$$

Denote $t = -q(\alpha + \epsilon)$. It holds that

$$\mu(B(x, r_n))^q (2r_n)^t > 2^t.$$

It results, as previously, that $\mathcal{H}_\mu^{q,t}(\{x\}) > 2^t$, which yields that

$$b_\mu(q) \geq t = -q(\alpha + \epsilon).$$

By letting $\epsilon \to 0$, we end up with a contradiction.
The same result can be proved by considering B_μ instead of b_μ. $\qquad\square$

We are now able to introduce, and prove the fundamental result in this chapter, which deals with the multifractal formalism for measures due to [Olsen (1995)]. Such a result applies the well known Borel-Cantelli lemma, and the large deviation formalism. We thus recall the two results before stating the fundamental theorem. A general case due to the mixed multifractal formalism for measures has been developed in [Menceur, Ben Mabrouk and Betina (2016); Menceur and Ben Mabrouk (2019)], where a generalized large deviation formalism has been established, for the case of non necessary Gibbs measures, and for a mixed multifractal analysis.

Lemma 8.2. *Borel-Cantelli Lemma.* *Let* $(\Omega, \mathcal{A}, \mathbb{P})$ *be a probability space. For all* $\mathcal{A}_n \in \mathcal{A}$, *we have*

$$\sum_{n \geq 1} \mathbb{P}(A_n) < \infty \Longrightarrow \mathbb{P}(\overline{\lim} A_n) = 0.$$

Theorem 8.2. Large deviations formalism. *Let $(W_n)_n$ be a sequence of random variables on a probability space $(\Omega, \mathcal{A}, \mathbb{P})$, and $(a_n)_n \subset]0, +\infty[$ with $\lim\limits_{n \to +\infty} a_n = \infty$. Let, for $n \in \mathbb{N}$,*

$$C_n(t) = \frac{1}{a_n} \log\Big(E(\exp(tW_n)) \Big),$$

and assume that $C_n(t)$ is finite for all n, and p, and that

$$C(t) = \lim_{n \to +\infty} C_n(t),$$

exists, and finite for all t. Then,

(1) *The function C is convex.*
(2) *If for $t \in \mathbb{R}$, $C'_-(t) \leq C'_+(t) < \alpha$, then*

$$\limsup_{n \to +\infty} \frac{1}{a_n} \log\left(e^{-a_n C(t)} E\left(\exp(tW_n) 1_{\{\frac{W_n}{a_n} \geq \alpha\}} \right) \right) < 0.$$

(3) *If $\sum\limits_n e^{-\epsilon a_n}$ is finite for all $\epsilon > 0$, then*

$$\limsup_{n \to +\infty} \frac{W_n}{a_n} \leq C'_+(0), \quad \mathbb{P} \ a.s.$$

(4) *If for $t \in \mathbb{R}$, $\alpha < C'_-(t) \leq C'_+(t)$,*

$$\limsup_{n \to +\infty} \frac{1}{a_n} \log\left(e^{-a_n C(t)} E\left(\exp(tW_n) 1_{\{\frac{W_n}{a_n} \leq \alpha\}} \right) \right) < 0.$$

(5) *Whenever $\sum\limits_n e^{-\epsilon a_n}$ is finite, for all $\epsilon > 0$, then*

$$C'_-(0) \leq \limsup_{n \to +\infty} \frac{W_n}{a_n}, \quad \mathbb{P} \ a.s.$$

Proof. (1) It is sufficient to prove that the function C_n is convex. Indeed, this function is twice differentiable, and

$$C_n"(t) = \frac{\left(\int_\Omega W_n^2 e^{tW_n} d\mathbb{P} \right)\left(\int_\Omega e^{tW_n} d\mathbb{P} \right) - \left(\int_\Omega W_n e^{tW_n} d\mathbb{P} \right)^2}{\left(\int_\Omega e^{tW_n} d\mathbb{P} \right)^2}.$$

The Cauchy-Schwartz's inequality applied to the functions

$$f = W_n \exp(\tfrac{1}{2} tW_n), \quad \text{and} \quad g = \exp(\tfrac{1}{2} tW_n)$$

permits to affirm that C_n is convex.

(2) As C is convex, there exists $\delta > 0$, such that,

$$C(t) + \alpha\delta - C(t + \delta) > 0.$$

Therefore,

$$\frac{1}{a_n} \log\left[e^{-a_n C(t)} \mathbb{E}\left(\exp(tW_n) 1_{\{\frac{W_n}{a_n} \geq \alpha\}} \right) \right]$$

$$= \frac{1}{a_n} \log\left[e^{-a_n C(t)} \int_{\{\frac{W_n}{a_n} \geq \alpha\}} e^{tW_n} d\mathbb{P} \right]$$

$$= \frac{1}{a_n} \log\left[e^{-a_n(C(t)+\alpha\delta)} \int_{\{\frac{W_n}{a_n} \geq \alpha\}} e^{tW_n + a_n\alpha\delta} d\mathbb{P} \right]$$

$$\leq \frac{1}{a_n} \log\left[e^{-a_n(C(t)+\alpha\delta)} \int_{\{\frac{W_n}{a_n} \geq \alpha\}} e^{(t+\delta)W_n} d\mathbb{P} \right]$$

$$\leq \frac{1}{a_n} \log\left[e^{-a_n(C(t)+\alpha\delta)} \mathbb{E}(\exp((t+\delta)W_n)) \right]$$

$$= \frac{1}{a_n} \log\left[e^{-a_n(C(t)+\alpha\delta - C_n(t+\delta))} \right]$$

$$= -(C(t) + \alpha\delta - C_n(t+\delta)).$$

By taking the upper limit we get the result.

(3) For $n, m \in \mathbb{N}$, consider the set

$$T_{n,m} = \{\frac{W_n}{a_n} \geq C'_+(0) + \frac{1}{m}\}.$$

Take next in assertion **(2)**, $t = 0$, and $\alpha = C'_+(0) + \frac{1}{m}$. We obtain

$$\limsup_n \frac{1}{a_n} \log \mathbb{P}(T_{n,m}) < 0.$$

Hence, there exists $\epsilon > 0$, and $N \in \mathbb{N}$, such that,

$$\limsup_n \frac{1}{a_n} \log \mathbb{P}(T_{n,m}) < -\epsilon, \ \forall n \geq N,$$

which leads to

$$\mathbb{P}(T_{n,m}) < e^{-\epsilon a_n}.$$

Consequently, the series $\sum_n \mathbb{P}(T_{n,m})$ is convergent. Borel-Cantelli Lemma yields that

$$\mathbb{P}(\limsup_n T_{n,m}) = 0, \ \forall m.$$

As a results,

$$\mathbb{P}\left(\limsup_n \frac{W_n}{a_n} > C'_+(0)\right) = \mathbb{P}(\bigcup_m \limsup_n T_{n,m}) = 0.$$

Consequently,

$$\limsup_n \frac{W_n}{a_n} \leq C'_+(0).$$

□

Theorem 8.3. *Let μ be a Borel probability measure on $\mathbb{R}^d$, q a fixed real number. Let $t_q \in \mathbb{R}$, r_q, $\underline{K}_q, \overline{K}_q \in]0, \infty[$, ν_q be a Borel probability measure on $Support(\mu)$, and $\varphi_q : \mathbb{R}_+ \to \mathbb{R}$ a function. Let $(r_{q,n})_n \subset]0, 1[$ be, such that,*

$$r_{q,n} \downarrow 0, \quad \frac{\log r_{q,n+1}}{\log r_{q,n}} \to 1, \quad and \quad \sum_n r_{q,n}^{\epsilon} < \infty \; \forall \epsilon > 0.$$

Assume next that
$\mathcal{H}_1$) $\forall \, x \in Support(\mu)$, and $r \in]0, r_q[$,

$$\underline{K}_q \leq \frac{\nu_q(B(x,r))}{\Big(\mu(B(x,r))\Big)^q (2r)^{t_q} \exp(\varphi_q(r))} \leq \overline{K}_q.$$

$\mathcal{H}_2$) $\varphi_q(r) = o(\log r)$, as $r \to 0$.
$\mathcal{H}_3$) $C_n(p) = \dfrac{1}{-\log r_{q,n}} \log\left(\int_{Support(\mu)} \Big(\mu(B(x,r_{q,n}))\Big)^p d\nu_q(x)\right)$ *exists,*
and finite, for all $p, n \in \mathbb{N}$.
$\mathcal{H}_4$) $C(p) = \lim\limits_{n \to +\infty} C_n(p)$ *exists, and finite for all $p \in \mathbb{R}$.*
Then, the following assertions hold.
(i) $dim(\underline{X}_{-C'_+(0)} \cap \overline{X}^{-C'_-(0)}) \geq$

$$\begin{cases} -C'_-(0)q + \Lambda_\mu(q) \geq -C'_-(0)q + B_\mu(q) \geq -C'_-(0)q + b_\mu(q) \,, \, q \leq 0 \\ -C'_+(0)q + \Lambda_\mu(q) \geq -C'_+(0)q + B_\mu(q) \geq -C'_+(0)q + b_\mu(q) \,, \, q \geq 0 \end{cases}$$

(ii) *If moreover C is differentiable at 0, then*

$$f_\mu(-C'(0)) = b_\mu^*(-C'(0)) = B_\mu^*(-C'(0)) = \Lambda_\mu^*(-C'(0)).$$

Proof. The important point of the proof is to show that the measure ν_q is supported by the set $\underline{X}_{-C'_+(0)} \cap \overline{X}^{-C'_-(0)}$. This follows in fact from the

Large Deviation Theorem. Indeed, denote $t = t_q$, $\underline{K} = \underline{K}_q$, $\overline{K} = \overline{K}_q$, $\varphi = \varphi_q$, $\nu = \nu_q$, and $r_n = r_{q,n}$. For $x \in Support(\mu)$, we assume that

$$\underline{\alpha}_\mu(x, r_n) = \liminf_n \frac{\log\left[\mu(B(x, r_n))\right]}{\log r_n},$$

and

$$\overline{\alpha}_\mu(x, r_n) = \limsup_n \frac{\log\left[\mu(B(x, r_n))\right]}{\log r_n}.$$

(i) We shall show that

$$b_\mu(q) = B_\mu(q) = \Lambda_\mu(q) = t.$$

Due to Proposition 8.8, it suffices to prove that

$$t \leq b_\mu(q), \quad \text{and} \quad \Lambda_\mu(q) \leq t.$$

Without loss the generality, we may assume that $r_q < 1$. Let $\epsilon > 0$, and $0 < \delta_\epsilon < r_q$ be, such that,

$$\left|\frac{\varphi(r)}{\log r}\right| < \epsilon, \quad \forall r; \ 0 < r < \delta_\epsilon.$$

Consider $\left(B(x_i, r_i)\right)_i$ a δ-packing of $Support(\mu)$. Due to the hypothesis $\mathcal{H}_1$, we obtain

$$2^{-\epsilon}\underline{K}\left(\mu(B(x_i, r_i))\right)^q (2r_i)^{t+\epsilon} \leq \underline{K}\left(\mu(B(x_i, r_i))\right)^q (2r_i)^t e^{\varphi(r_i)}$$
$$\leq \nu\left(B(x_i, r_i)\right).$$

It holds that

$$\sum_i \left(\mu(B(x_i, r_i))\right)^q (2r_i)^{t+\epsilon} \leq \frac{2^\epsilon}{\underline{K}} \sum_i \nu\left(B(x_i, r_i)\right)$$
$$= \frac{2^\epsilon}{\underline{K}} \nu\left(\bigcup_i B(x_i, r_i)\right)$$
$$\leq \frac{2^\epsilon}{\underline{K}}.$$

Hence,

$$\overline{\mathcal{P}}_{\mu,\delta}^{q,t+\epsilon}(Support(\mu)) \leq \frac{2^\epsilon}{\underline{K}}, \quad \forall \ 0 < \delta < \delta_\epsilon.$$

Consequently,

$$\Lambda_\mu(q) \leq t + \epsilon, \quad \forall \epsilon > 0.$$

This yields, finally, that,

$$\Lambda_\mu(q) \leq t.$$

Consider now that ϵ-covering $\Big(B(x_i, r_i)\Big)_i$ of $Support(\mu)$. The hypothesis $\mathcal{H}_1$ yields that

$$2^\epsilon \overline{K}\Big(\mu(B(x_i, r_i))\Big)^q (2r_i)^{t-\epsilon} \geq \overline{K}\Big(\mu(B(x_i, r_i))\Big)^q (2r_i)^t e^{\varphi(r_i)}$$
$$\geq \nu\Big(B(x_i, r_i)\Big).$$

As a result,

$$\sum_i \Big(\mu(B(x_i, r_i))\Big)^q (2r_i)^{t-\epsilon} \geq \frac{2^{-\epsilon}}{\overline{K}} \sum_i \nu\Big(B(x_i, r_i)\Big) \geq \frac{2^{-\epsilon}}{\overline{K}}.$$

Therefore,

$$\overline{\mathcal{H}}_{\mu,\delta}^{q,t-\epsilon}(Support(\mu)) \geq \frac{2^{-\epsilon}}{\overline{K}}, \quad \forall\, 0 < \delta < \delta_\epsilon,$$

which leads to

$$b_\mu(q) \geq t - \epsilon, \quad \forall \epsilon > 0.$$

Hence,

$$b_\mu(q) \geq t.$$

Consider now the set

$$M = \Big\{x \in Support(\mu);\ -C'_+(0) \leq \underline{\alpha}_\mu(x, r_n) \leq \overline{\alpha}_\mu(x, r_n) \leq -C'_-(0)\Big\}.$$

We will prove that

$$M = \underline{X}_{-C'_+(0)} \cap \overline{X}_{-C'_-(0)}.$$

So, let $n \in \mathbb{N}$, and $r > 0$ be, such that, $r_{n+1} \leq r \leq r_n$. We have

$$\frac{\log r_n}{\log r_{n+1}} \frac{\log[\mu(B(x, r_n))]}{\log r_n} \leq \frac{\log[\mu(B(x, r))]}{\log r} \leq \frac{\log r_{n+1}}{\log r_n} \frac{\log[\mu(B(x, r_{n+1}))]}{\log r_{n+1}}.$$

This gives that for all $x \in Support(\mu)$, We have

$$\underline{\alpha}_\mu(x) = \underline{\alpha}_\mu(x, r_n), \quad \text{and} \quad \overline{\alpha}_\mu(x) = \overline{\alpha}_\mu(x, r_n).$$

Then the equality

$$M = \underline{X}_{-C'_+(0)} \cap \overline{X}_{-C'_-(0)}.$$

We will apply now the Large deviation Theorem. So let's take

$$\Omega = Support(\mu), \quad \mathcal{A} = \mathcal{B}(Support(\mu)),$$

$$\mathbb{P} = \mu, \quad W_n(x) = \log(\mu(B(x, r_n))), \quad \text{and} \quad a_n = -\log r_n.$$

It holds that

$$-C'_+(0) \le \underline{a}_\mu(x, r_n) \le \overline{a}_\mu(x, r_n) \le -C'_-(0) \quad \nu. \, p.s.$$

Consequently, the measure ν is supported by M. If $x \in Support(\mu)$, and $0 < r < r_n$, the hypothesis $\mathcal{H}_1$ yields that, for all $x \in M$,

$$\underline{a}_\mu(x) \ge \begin{cases} -qC'_-(0) + t \text{ for } q \le 0, \\ -qC'_+(0) + t \text{ for } q \ge 0. \end{cases}$$

As $\nu(M) = 1$, we conclude due to Billingsley's Theorem, that

$$dim \, M \ge \begin{cases} -qC'_-(0) + t \text{ for } q \le 0, \\ -qC'_+(0) + t \text{ for } q \ge 0. \end{cases}$$

(2) If C is differentiable at 0, assertion **(1)** leads to

$$dim \, M \ge -qC'(0) + t \ge \Lambda^*_\mu(-C'(0)) \ge B^*_\mu(-C'(0)) \ge b^*_\mu(-C'(0)).$$

On the other hand, $\nu(M) = 1$. The set M is then non empty. Due to the Proposition 8.8, we obtain

$$-qC'(0) + t \ge 0.$$

Consequently,

$$dim \, M \le -qC'(0) + t, \, \forall q \in \mathbb{R}.$$

By taking the lower bound on q, we obtain

$$dim \, M \le b^*_\mu(-C'(0)) \le B^*_\mu(-C'(0)) \le \Lambda^*_\mu(-C'(0)).$$

$\square$

Corollary 8.1. *Assume that the hypotheses $\mathcal{H}_1$, $\mathcal{H}_2$, $\mathcal{H}_3$, and $\mathcal{H}_4$ of Theorem 8.3 hold, and moreover C is differentiable at 0, then*

$$f_\mu(-C'(0)) = b^*_\mu(-C'(0)) = B^*_\mu(-C'(0)) = \Lambda^*_\mu(-C'(0)).$$

Proof. The hypothesis $\mathcal{H}_1$ is true for all $q \in \mathbb{R}$. Therefore, for all $q < 0$, $x \in Support(\mu)$, and r, $0 < r < r_q$, we have

$$1 \le \frac{\nu(B(x, 2r))}{\nu(B(x, r))} \le \frac{\overline{K}}{\underline{K}} 2^t \exp(\varphi(2r) - \varphi(r)) \left[\frac{\mu(B(x, 2r))}{\mu(B(x, r))} \right]^q.$$

Therefore, for all $x \in Support(\mu)$, and r, $0 < r < r_q$

$$\frac{\mu(B(x,2r))}{\mu(B(x,r))} < \beta,$$

where

$$\beta = \left[\frac{\underline{K}}{\overline{K}}2^{-t}\right]^{1/q}\exp(\frac{1}{q}[\varphi(r) - \varphi(2r)]).$$

Let $(B(x_{ij}, r_n))_{1 \leq i \leq \xi, j}$ be the ξ families defined in Besicovitch's covering Theorem extracted from the family $(B(x_i, r_n))_i$. We will proceed by steps.

Step 1: $p + q - 1 \geq 0$.

If $p \geq 0$, we have

$$\underline{K}(2r_n)^t \exp(\varphi(r_n)) \int_{B(x_{ij}, r_n)} \left[\mu(B(y, r_n))\right]^{p+q-1} d\mu(y)$$

$$\leq \underline{K}(2r_n)^t \exp(\varphi(r_n)) \int_{B(x_{ij}, r_n)} \left[\mu(B(x_{ij}, 2r_n))\right]^{p+q-1} d\mu(y)$$

$$\leq \underline{K}(2r_n)^t \exp(\varphi(r_n)) \left[\mu(B(x_{ij}, 2r_n))\right]^{p+q}$$

$$\leq \beta^{p+q} \left[\mu(B(x_{ij}, r_n))\right]^p \nu(B(x_{ij}, r_n))$$

$$\leq \beta^{p+q} \int_{B(x_{ij}, r_n)} \left[\mu(B(x_{ij}, r_n))\right]^p d\nu(y)$$

$$\leq \beta^{p+q} \int_{B(x_{ij}, r_n)} \left[\mu(B(y, 2r_n))\right]^p d\nu(y),$$

which leads to

$$(p + q - 1)\overline{I}_\mu^{p+q-1} \leq C_q(p) + t_q.$$

Conversely,

$$\overline{K}(2r_n)^t \exp(\varphi(r_n)) \int_{B(x_{ij}, r_n)} \left[\mu(B(y, 2r_n))\right]^{p+q-1} d\mu(y)$$

$$\geq \overline{K}(2r_n)^t \exp(\varphi(r_n)) \int_{B(x_{ij}, r_n)} \left[\mu(B(x_{ij}, r_n))\right]^{p+q-1} d\mu(y)$$

$$\geq \overline{K}(2r_n)^t \exp(\varphi(r_n)) \left[\mu(B(x_{ij}, r_n))\right]^{p+q}$$

$$\geq \beta^{-p} \int_{B(x_{ij}, r_n)} \left[\mu(B(y, r_n))\right]^p d\nu(y).$$

Consequently,

$$(p + q - 1)\underline{I}_\mu^{p+q-1} \geq C_q(p) + t_q.$$

We suppose now that $p \leq 0$. We have

$$\underline{K}(2r_n)^t \exp(\varphi(r_n)) \int_{B(x_{ij},r_n)} \left[\mu(B(y,r_n))\right]^{p+q-1} d\mu(y)$$

$$\leq \underline{K}(2r_n)^t \exp(\varphi(r_n)) \int_{B(x_{ij},r_n)} \left[\mu(B(x_{ij},2r_n))\right]^{p+q-1} d\mu(y)$$

$$\leq \underline{K}(2r_n)^t \exp(\varphi(r_n)) \left[\mu(B(x_{ij},2r_n))\right]^{p+q}$$

$$\leq \beta^{p+q} \left[\mu(B(x_{ij},r_n))\right]^p \nu(B(x_{ij},r_n))$$

$$\leq \beta^{p+q} \int_{B(x_{ij},r_n)} \left[\mu(B(x_{ij},r_n))\right]^p d\nu(y)$$

$$\leq \beta^p \int_{B(x_{ij},r_n)} \left[\mu(B(y,r_n))\right]^p d\nu(y).$$

As a result,

$$(p+q-1)\overline{I}_\mu^{p+q-1} \leq C_q(p) + t_q.$$

Conversely,

$$\overline{K}(2r_n)^t \exp(\varphi(r_n)) \int_{B(x_{ij},r_n)} \left[\mu(B(y,2r_n))\right]^{p+q-1} d\mu(y)$$

$$\geq \overline{K}(2r_n)^t \exp(\varphi(r_n)) \int_{B(x_{ij},r_n)} \left[\mu(B(x_{ij},r_n))\right]^{p+q-1} d\mu(y)$$

$$\geq \overline{K}(2r_n)^t \exp(\varphi(r_n)) \left[\mu(B(x_{ij},r_n))\right]^{p+q}$$

$$\geq \beta^{-p} \int_{B(x_{ij},r_n)} \left[\mu(B(y,r_n))\right]^p d\nu(y).$$

Therefore,

$$(p+q-1)\underline{I}_\mu^{p+q-1} \geq C_q(p) + t_q.$$

Step 2: $p+q-1 \leq 0$.

For $p \leq 0$, We have

$$\underline{K}(2r_n)^t \exp(\varphi(r_n)) \int_{B(x_{ij},r_n)} \left[\mu(B(y,2r_n))\right]^{p+q-1} d\mu(y)$$

$$\leq \underline{K}(2r_n)^t \exp(\varphi(r_n)) \int_{B(x_{ij},r_n)} \left[\mu(B(x_{ij},r_n))\right]^{p+q-1} d\mu(y)$$

$$\leq \underline{K}(2r_n)^t \exp(\varphi(r_n)) \left[\mu(B(x_{ij},r_n))\right]^{p+q}$$

$$\leq \left[\mu(B(x_{ij},r_n))\right]^{p} \nu(B(x_{ij},r_n))$$

$$\leq \int_{B(x_{ij},r_n)} \left[\mu(B(x_{ij},r_n))\right]^{p} d\nu(y)$$

$$\leq \beta^{-p} \int_{B(x_{ij},r_n)} \left[\mu(B(y,r_n))\right]^{p} d\nu(y).$$

This yields that

$$(p+q-1)\overline{I}_\mu^{p+q-1} \leq C_q(p) + t_q.$$

Conversely,

$$\overline{K}(2r_n)^t \exp(\varphi(r_n)) \int_{B(x_{ij},r_n)} \left[\mu(B(y,r_n))\right]^{p+q-1} d\mu(y)$$

$$\geq \overline{K}(2r_n)^t \exp(\varphi(r_n))\beta^{p+q-1} \int_{B(x_{ij},r_n)} \left[\mu(B(x_{ij},r_n))\right]^{p+q-1} d\mu(y)$$

$$\geq \overline{K}(2r_n)^t \exp(\varphi(r_n))\beta^{p+q-1} \left[\mu(B(x_{ij},r_n))\right]^{p+q}$$

$$\geq \beta^{p+q-1} \int_{B(x_{ij},r_n)} \left[\mu(B(y,2r_n))\right]^{p} d\nu(y).$$

Consequently,

$$(p+q-1)\underline{I}_\mu^{p+q-1} \geq C_q(p) + t_q.$$

We suppose now that $p \geq 0$. We have

$$\underline{K}(2r_n)^t \exp(\varphi(r_n)) \int_{B(x_{ij}, r_n)} \left[\mu(B(y, 2r_n)) \right]^{p+q-1} d\mu(y)$$

$$\leq \underline{K}(2r_n)^t \exp(\varphi(r_n)) \int_{B(x_{ij}, r_n)} \left[\mu(B(x_{ij}, r_n)) \right]^{p+q-1} d\mu(y)$$

$$\leq \underline{K}(2r_n)^t \exp(\varphi(r_n)) \left[\mu(B(x_{ij}, r_n)) \right]^{p+q}$$

$$\leq \left[\mu(B(x_{ij}, r_n)) \right]^{p} \nu(B(x_{ij}, r_n))$$

$$\leq \int_{B(x_{ij}, r_n)} \left[\mu(B(x_{ij}, r_n)) \right]^{p} d\nu(y)$$

$$\leq \int_{B(x_{ij}, r_n)} \left[\mu(B(y, 2r_n)) \right]^{p} d\nu(y).$$

As a result,

$$(p + q - 1)\overline{I}_\mu^{p+q-1} \leq C_q(p) + t_q.$$

Conversely,

$$\overline{K}(2r_n)^t \exp(\varphi(r_n)) \int_{B(x_{ij}, r_n)} \left[\mu(B(y, r_n)) \right]^{p+q-1} d\mu(y)$$

$$\geq \overline{K}(2r_n)^t \exp(\varphi(r_n)) \beta^{p+q-1} \int_{B(x_{ij}, r_n)} \left[\mu(B(x_{ij}, 2r_n)) \right]^{p+q-1} d\mu(y)$$

$$\geq \overline{K}(2r_n)^t \exp(\varphi(r_n)) \beta^{p+q-1} \left[\mu(B(x_{ij}, r_n)) \right]^{p+q}$$

$$\geq \beta^{q-1} \int_{B(x_{ij}, r_n)} \left[\mu(B(y, r_n)) \right]^{p} d\nu(y).$$

Therefore,

$$(p + q - 1)\underline{I}_\mu^{p+q-1} \geq C_q(p) + t_q.$$

We deduce that for all $p, q \in \mathbb{R}$,

$$(p + q - 1)I_\mu^{p+q-1} = C_q(p) + t_q.$$

As a result,

$$(p + q - 1)I_\mu^{p+q-1} = C_\mu^{p+q}(Support(\mu)) = \Lambda_\mu(p + q).$$

Consequently,

$$C_q(p) = \Lambda_\mu(p + q) - \Lambda_\mu(p).$$

Thus, if Λ_μ is differentiable at q, C_q will be differentiable at 0, and

$$C_q'(0) = \Lambda_\mu'(q).$$

The Large Deviation theorem implies that

$$\alpha_\mu(x) = -C_q'(0) \; ; \; \nu_q, \text{ for almost every } x \in support(\mu).$$

Finally,

$$\alpha_\mu(x) = -\Lambda_\mu'(q).$$

$\square$

Corollary 8.2. *Assume that the hypotheses* $\mathcal{H}_1$, $\mathcal{H}_2$, $\mathcal{H}_3$, *and* $\mathcal{H}_4$ *of Theorem 8.3 hold for all* $q \in \mathbb{R}$. *The following assertions hold.*
(i) *If* B_μ *is differentiable at* q, *then* $\alpha_\mu = -B_\mu$, ν_q, *a.s, and*

$$\{-B_\mu'(q); \; B_\mu' \text{ exists} \} \subseteq \alpha_\mu(support(\mu)).$$

(ii) $f_\mu = B_\mu^*$ *on* $\{ -B_\mu'(q); \; B_\mu' \text{ exists}\}$.

Proof. Let q, such that, $\Lambda_\mu'(q)$ exists. So $C_q'(0)$ exists. Corollary 8.1 leads to

$$f_\mu(-C'(0)) = \Lambda_\mu^*(-C'(0)).$$

$\square$

8.4   Exercises for Chapter 8

Exercise 1.
Develop a proof of Lemma 8.2 (the Borel-Cantelli Lemma).

Exercise 2.
Let Ω be the set of open bounded intervals in $\mathbb{R}$, and ℓ be the set function written on the

$$\ell(a, b) = (b - a)^s,$$

for some real number parameter s. Let next the set function

$$\mathcal{H}(A) = inf\{\sum_k \ell_i(I_k); \; A \subset \cup_k I_k, I_k \in \Omega, \forall k\}.$$

(1) For what values of s the set function $\mathcal{H}$ is an outer measure on Ω?
(2) Evaluate the behavior of $\ell_s(x - r, x + r)$, as r goes to 0.

Exercise 3.

Let

$$h(r) = r^\alpha (\log|\log r|)^\beta,$$

where α, β are constants. Consider

$$\mathcal{H}^h(E) = \lim_{\varepsilon \searrow 0} \inf \left\{ \sum_i h(B(x_i, r_i)); \ x_i \in E \subset \bigcup_i B(x_i, r_i), \ r_i < \varepsilon \right\}.$$

(1) Study the possible Hausdorff measure properties of $\mathcal{H}$.
(2) Study the same problem with packings instead of coverings.

Exercise 4.

Let

$$h(r) = r^\alpha (\log|\log r|)^\beta,$$

where α, β are constants. For a Borel probability measure μ on R, Consider for $q \in \mathbb{R}$ the set function,

$$\mathcal{H}_\mu^{h,q}(E) = \lim_{\varepsilon \searrow 0} \inf \left\{ \sum_i \mu(B(x_i, r_i))^q h(B(x_i, r_i)); \ x_i \in E \subset \bigcup_i B(x_i, r_i), \ r_i < \varepsilon \right\}.$$

1) Study the possible Hausdorff measure properties of $\mathcal{H}_\mu^{h,q}$.
2) Study the same problem with packings instead of coverings.

Exercise 5.

Consider the function

$$f(t) = -\frac{\gamma(\log t)^{\gamma-1}}{t} e^{-(|\log t|)^\delta} X_{]0,1[}(t),$$

where $\gamma > 0$ is a constant, and the measure $\mu(t) = f(t)dt$.
(1) Compute $\alpha_\mu(0)$.
(2) Show that for $x \in (0,1)$,

$$\log \mu(B(x,r)) = -(\log x)^2 - \frac{2\log x}{x} r + r\varepsilon(r), \quad r \to 0.$$

(3) Deduce $\alpha_\mu(x)$, for $x \in [0,1]$.
(4) Let $\varphi(r) = \log r(|\log r|)^{\gamma-1}$, and

$$\nu_{q,t}(B(x,r)) = (\mu(B(x,r)))^q e^{t\varphi(r)}.$$

Show that

$$\frac{\nu_{q,t}(B(x,2r))}{\nu_{q,t}(B(x,r))} \sim \left[\frac{\mu(B(x,2r))}{\mu(B(x,r))} \right]^q e^{\gamma(\log 2)|\log r|^{\gamma-1}}, \quad \text{as } r \to 0.$$

(5) Deduce that

$$\frac{\nu_{q,t}(B(x,2r))}{\nu_{q,t}(B(x,r))} \sim C_{q,\mu} e^{\gamma(\log 2)|\log r|^{\gamma-1}} \to \infty \text{ (or } 0) \text{ as } r \to 0.$$

Exercise 6.

Let $\varphi : \mathbb{R}_+ \to \mathbb{R}$ be, such that,

$$\varphi \text{ is non-decreasing, and } \varphi(r) < 0, \text{ for } r \text{ small enough,}$$

and consider, for a metric space (X, d), the function

$$h_{q,t}(r) = \mu(B(x,r))^q e^{t\varphi(r)}, \quad r > 0, \ x \in X, \ q \in \mathbb{R}.$$

Consider the quantity

$$\overline{\mathcal{H}}^{q,t}_{\mu,\varphi,\epsilon}(E) = \inf\{ \sum_i h_{q,t}(r) \},$$

where the inf is taken over the set of all centered ϵ-coverings of E, and for the empty set, $\overline{\mathcal{H}}^{q,t}_{\mu,\epsilon}(\emptyset) = 0$. Denote next

$$\overline{\mathcal{H}}^{q,t}_{\mu,\varphi}(E) = \lim_{\epsilon \downarrow 0} \overline{\mathcal{H}}^{q,t}_{\mu,\varphi,\epsilon}(E) = \sup_{\delta > 0} \overline{\mathcal{H}}^{q,t}_{\mu,\varphi,\epsilon}(E),$$

and finally,

$$\mathcal{H}^{q,t}_{\mu,\varphi}(E) = \sup_{F \subseteq E} \overline{\mathcal{H}}^{q,t}_{\mu,\varphi}(F).$$

(1) Show that $\mathcal{H}^{q,t}_{\mu,\varphi}$ is an outer metric measure on $\mathbb{R}^d$, for which, Borel sets are measurable.

(2) Develop the multifractal analysis of $\mathcal{H}^{q,t}_{\mu,\varphi}$.

Exercise 7.

Let $\varphi : \mathbb{R}_+ \to \mathbb{R}$ be, such that,

$$\varphi \text{ is non-decreasing, and } \varphi(r) < 0, \text{ for } r \text{ small enough,}$$

and consider for a metric space (X, d) the function

$$h_{q,t}(r) = \mu(B(x,r))^q e^{t\varphi(r)}, \quad r > 0, \ x \in X, \ q \in \mathbb{R}.$$

Consider the quantity

$$\overline{\mathcal{P}}^{q,t}_{\mu,\varphi,\epsilon}(E) = \sup\{ \sum_i (\mu(B(x_i, r_i)))^q e^{t\varphi(r_i)} \},$$

where the sup is taken over the set of all centered ϵ-packings of E. For the empty set, we set, as usual, $\overline{\mathcal{P}}^{q,t}_{\mu,\varphi,\epsilon}(\emptyset) = 0$. Next, we consider the limit as $\epsilon \downarrow 0$,

$$\overline{\mathcal{P}}^{q,t}_{\mu,\varphi}(E) = \lim_{\epsilon \downarrow 0} \overline{\mathcal{P}}^{q,t}_{\mu,\varphi,\epsilon}(E) = \inf_{\epsilon > 0} \overline{\mathcal{P}}^{q,t}_{\mu,\varphi,\epsilon}(E),$$

and finally,

$$\mathcal{P}_{\mu,\varphi}^{q,t}(E) = \inf_{E \subseteq \cup_i E_i} \sum_i \overline{\mathcal{P}}_{\mu,\varphi}^{q,t}(E_i).$$

(1) Show that $\mathcal{P}_{\mu,\varphi}^{q,t}$ is an outer metric measure on $\mathbb{R}^d$, for which, Borel sets are measurable.

(2) Develop the multifractal analysis of $\mathcal{P}_{\mu,\varphi}^{q,t}$.

Exercise 8.

Let $\varphi : \mathbb{R}_+ \to \mathbb{R}$ be, such that,

$$\varphi \text{ is non-decreasing, and } \varphi(r) < 0, \text{ for } r \text{ small enough.}$$

Let μ be a Borel probability measure satisfying the assumption

$$\mathcal{A}_{\mu,\varphi}(a, \alpha) = \limsup_{r \to 0} \left(\sup_{x \in \mathcal{S}_\mu} e^{\alpha \varphi(r)} \frac{\mu(B(x, ar))}{\mu(B(x, r))} \right) < \infty,$$

for some $a > 1$, and for all $\alpha > 0$, where $\mathcal{S}_\mu$ is the support of μ. Assume further that

$$\mathcal{H}_{\mu,\varphi}^{q, \Lambda_{\mu,\varphi}(q)}(\mathcal{S}_\mu) > 0,$$

for some $q \in \mathbb{R}_+^k$. Show that there exists a Borel probability measure ν supported by $\mathcal{S}_\mu$, such that,

$$\nu(B(x, r)) \leq \overline{K} \Big(\mu(B(x, r)) \Big)^q e^{\Lambda_{\mu,\varphi}(q) \varphi(r)}; \quad \forall x \in \mathcal{S}_\mu, \ \forall 0 < r << 1.$$

Exercise 9.

Let $\varphi : \mathbb{R}_+ \to \mathbb{R}$ be, such that,

$$\varphi \text{ is non-decreasing, and } \varphi(r) < 0, \text{ for } r \text{ small enough.}$$

Let μ be a Borel probability measure satisfying the assumption

$$\mathcal{A}_{\mu,\varphi}(a, \alpha) = \limsup_{r \to 0} \left(\sup_{x \in \mathcal{S}_\mu} e^{\alpha \varphi(r)} \frac{\mu(B(x, ar))}{\mu(B(x, r))} \right) < \infty,$$

for some $a > 1$, and for all $\alpha > 0$, where $\mathcal{S}_\mu$ is the support of μ. Assume further that

$$\mathcal{H}_{\mu,\varphi}^{q, B_{\mu,\varphi}(q)}(\mathcal{S}_\mu) > 0,$$

for some $q \in \mathbb{R}_+^k$. Show that there exists a Borel probability measure ν supported by $\mathcal{S}_\mu$, such that,

$$\nu(B(x, r)) \leq \overline{K} \Big(\mu(B(x, r)) \Big)^q e^{B_{\mu,\varphi}(q) \varphi(r)}; \quad \forall x \in \mathcal{S}_\mu, \ \forall 0 < r << 1.$$

Exercise 10.

Let $\varphi : \mathbb{R}_+ \to \mathbb{R}$ be, such that,

$$\varphi \text{ is non-decreasing, and } \varphi(r) < 0, \text{ for } r \text{ small enough.}$$

Let μ be a Borel probability measure on $\mathbb{R}^d$, and denote $\mathcal{S}_\mu$ its support.

(1) Assume that $\mathcal{H}_{\mu,\varphi}^{q,\Lambda_{\mu,\varphi}(q)}(\mathcal{S}_\mu) > 0$ for some $q \in \mathbb{R}_-^k$. Show that there exists a Borel probability measure ν supported by $\mathcal{S}_\mu$, such that,

$$\nu(B(x,r)) \leq \overline{K}\Big(\mu(B(x,r))\Big)^q e^{\Lambda_{\mu,\varphi}(q)\varphi(r)}; \quad \forall x \in S_\mu, \ \forall 0 < r << 1.$$

(2) Assume that $\mathcal{H}_{\mu,\varphi}^{q,B_{\mu,\varphi}(q)}(\mathcal{S}_\mu) > 0$ for some $q \in \mathbb{R}_-^k$. Show that there exists a Borel probability measure ν supported by $\mathcal{S}_\mu$, such that,

$$\nu(B(x,r)) \leq \overline{K}\Big(\mu(B(x,r))\Big)^q e^{B_{\mu,\varphi}(q)\varphi(r)}; \quad \forall x \in S_\mu, \ \forall 0 < r << 1.$$

Chapter 9

Some Applications

9.1 Introduction

In the present chapter we aim to discuss, and develop eventual applications of fractal analysis. In fact, such applications can not be limited, as fractals may be met everywhere in nature, and sciences. Therefore, we will try to present some simple, and concrete examples in both theoretical aspects, and experimental ones, and which will be tackle-able by readers in different levels, and different specialities using fractals.

Many things in nature, and life are fractal-like objects, and/or have some fractal-like properties. Everywhere around us, our natural, and naked vision permit to detect fractals, in trees, their leaves, electric bolts in the sky, clouds, blood network, vessels networks, stock prices, ... , etc.

The fractal, and multifractal analysis, and geometry yield mathematical, physical, and generally scientific tools to understand, and discover the exact laws behind these objects.

This geometry is nowadays widely, and rapidly growing up, especially, in applications in quasi all fields, such as, computer graphics, computer vision, video games, statistical control, climatology, oceanography, osteoporosis, cosmology, medical diagnostics, and also sociology, design, and arts.

Backgrounds on concrete applications, algorithms, discussions may be found in [Barnsley et al (1988); Briggs (1992); Dolotin and Morozov (2006); Flake (1998); Frame and Cohen (2015); Hudson and Mandelbrot (2005); Mandelbrot (1982, 1997, 1999, 2004, 2012); Mandelbrot and Hudson (2006); Novak (2004); Peitgen et al (1988, 1992a,b); Pesin and Climenhaga (2009); Pickover (1995); Rashid (2014); Scholz and Mandelbrot (1989); Stauffer and Stanley (1991)].

More specialized applications may be found in pure mathematics, where fractals appear in many contexts, such as, number theory, PDEs, ... etc, ([Alimohammady et al (2017); Baleanu et al (2014); Cattani (2020); Xiaojun et al (2013); Xu et al (2014); Yang, Cattani et al (2014); Yang, Cattani and Xie (2015); Yang, Machado et al (2017); Yang, Srivastava et al (2015); Yang, Zhang et al (2014a,b); Yan et al (2014); Zhang, Cattani and Yang (2015)]). Fractals appear also in dynamical systems [Gervais (2009); Pesin (1997); Pesin and Climenhaga (2009)], and turbulence modeling [Benzi et al (1984); Cattani and Pierro (2013); Cattani (2010b); Frisch and Parisi (1985); Mandelbrot (1974)]. Applications of fractals may be also found in prices modeling, markets, financial indices, [Azizieh (2002); Benaych-Georges (2009); Ben Mabrouk, Ben Abdallah and Hamrita (2011); Calvet and Fisher (2008); Fan et al (2019); Fernandez-Martinez et al (2019); Fillol (2005); Hudson and Mandelbrot (2005); Mandelbrot (1997); Mandelbrot and Hudson (2006); Walter (2001)]. Besides, fractals are also applied for modeling instruments, such as, fractal antenna ([Anguera et al (2020)], climate factors, and geophysical targets [Bozkus et al (2020); Chu (1999); Figueiredo et al (2014); Scholz and Mandelbrot (1989)]. In bio domains, such as, medical applications, fractals are nowadays famous models [Badea et al (2013); Castiglioni-Faini (2019); Cattani, Pierro and Altieri (2012); Karaca and Cattani (2017); Mauroy et al (2004); Sapoval and Filoche (2010); Weibel (1963)]. Finally, we may also find them in signal/image processing, such as, [Barnsley et al (1988); Cattani and Ciancio (2016); Chen, Cattani and Zhong (2014); Li et al (2014); Liu et al (2019); Vehel and Legrand (2004); Vehel and Vojak (1998)], traffic, arts, design, nature, ... etc, [Briggs (1992); Flake (1998); Li, Zhao and Cattani (2013); Mandelbrot (1982); Novak (2004); Pickover (1995); Scheuring and Riedi (1994); Wang et al (2014)].

9.2　Fractals in plants' nature

In the present section, we aim to illustrate some fractal-like objects in nature, provided with some eventual discussions about their laws, structure, and evolution. Readers interested may refer to many beautiful references, and thus, reproduce, re-develop these objects, and other different cases, [Barnsley et al (1988); Briggs (1992); Dolotin and Morozov (2006); Flake (1998); Frame and Cohen (2015); Hudson and Mandelbrot (2005); Mandelbrot (1982, 1999, 2004, 2012); Novak (2004); Peitgen et al (1988, 1992a,b); Pickover (1995); Rashid (2014); Scholz and Mandelbrot (1989); Stauffer and Stanley (1991)].

One of the visual natural fractals may be seen in trees. Every tree is composed of a self-similar type, and/or fractal-like object starting with a simple geometrical object, such as, a segment, which is transformed next by means of special transformations, which make the whole object looking like its pieces. Figure 9.1 illustrates the fractal structure, and/or the self-similar characteristics of the trees' ferns.

Fig. 9.1: A natural Barnsley fern.

These natural trees, and/or ferns may be modeled by fractals. The original idea is due to Barnsley, who proposed a mathematical model, which has been next proved to be one of the self-similar models described by means of the fractal analysis, and geometry. The mathematical model is based on the following system of transformations [Barnsley (2000); Barnsley et al (1988)].

$$S_1 \begin{pmatrix} x \\ y \end{pmatrix} = \begin{pmatrix} 0.00 & 0.00 \\ 0.00 & 0.16 \end{pmatrix} \begin{pmatrix} x \\ y \end{pmatrix},$$

$$S_2 \begin{pmatrix} x \\ y \end{pmatrix} = \begin{pmatrix} 0.85 & 0.04 \\ -0.04 & 0.85 \end{pmatrix} \begin{pmatrix} x \\ y \end{pmatrix} + \begin{pmatrix} 0.00 \\ 1.60 \end{pmatrix},$$

$$S_3 \begin{pmatrix} x \\ y \end{pmatrix} = \begin{pmatrix} 0.20 & -0.26 \\ 0.23 & 0.22 \end{pmatrix} \begin{pmatrix} x \\ y \end{pmatrix} + \begin{pmatrix} 0.00 \\ 1.60 \end{pmatrix},$$

$$S_4 \begin{pmatrix} x \\ y \end{pmatrix} = \begin{pmatrix} -0.15 & 0.28 \\ 0.26 & 0.24 \end{pmatrix} \begin{pmatrix} x \\ y \end{pmatrix} + \begin{pmatrix} 0.00 \\ 0.44 \end{pmatrix}.$$

The system of these contractive similarities admits the Barnsley fern as an attractor or as the unique non-empty invariant compact set $\mathbb{B}$, satisfying the invariance set equation

$$\mathbb{B} = S_1(\mathbb{B}) \cup S_2(\mathbb{B}) \cup S_3(\mathbb{B}) \cup S_4(\mathbb{B}).$$

The following algorithm permits to illustrate graphically the Barnsley fern. The detailed version may be found in [Mearns (2021)]. The numerical illustration of the Barnsley fern $\mathbb{B}$ is subject of Figure 9.2 below.

Algorithm 1: Barnsley numerical fern ([Mearns (2021)]).

Input: The similarities S_i, $1 \leq 1 \leq 4$;

Output: The numerical fern;

1 The initial point $I = [x; y] = [0; 0]$;

2 Iterations

3 Chose wit a probability p_1 the scheme

4 $T_1 = S_1(I)$; (This will display the first right branch of the fern).

5 Chose with a probability p_2 the scheme

6 $T_2 = S_2(T_1)$; (This will display the first left branch of the fern).

7 Chose with a probability p_3 the scheme

8 $T_3 = S_3(T_2)$;

9 Chose with a probability p_3 the scheme

10 $T_4 = S_4(T_3)$;

11 end iteration

Fig. 9.2: A numerical Barnsley fern.

9.3 Fractals in human body anatomy

In the human body anatomy, we discover that our bodies, cells, vessels, and many organs possess fractal structure, such as, human lung. The specific geometric fractal structure of this organ is one of the strong explanation of the possibility, and/or the ability of the thorax to enclose it. Indeed, the exchange of gasses, such as, the oxygen delivered to cells by the blood, and the carbon dioxide rejected is assured by the so-called alveoli. The total area of the lung of a normal adult is approximated by $100\ m^2$. This means that without a special distribution circuit, it can not be enclosed in the small volume of our thorax. It is thus the hierarchical or fractal structure of this tree which explains this amazing distribution circuit. Figure 9.3 shows the hierarchical, and fractal structure of the human lung. It show clearly the self-similar, and the scaling laws in such an organ.

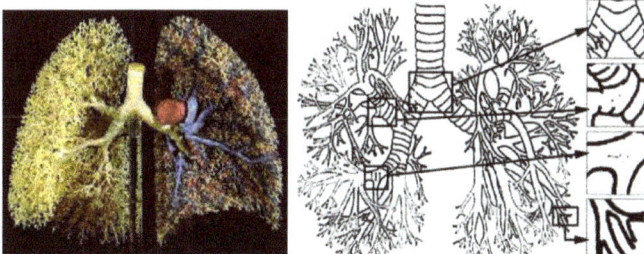

Fig. 9.3: Human lung.

To express more, and to understand the role, and the functionality of the lung, researchers, especially, in mathematics have focused on the study of the geometric structure, and thus many models have been developed to optimally simulate it. See for example [Mauroy et al (2004); Sapoval and Filoche (2010); Weibel (1963)]. The first fractal model in human lung simulation is due to Mandelbrot, and is illustrated in Figure 9.5. We will provide a general explicit mathematical construction of such model. Firstly we will develop the 2-dimensional case, which is more easier that the 3-dimensional one, and thus will permit to the readers a good comprehension of how to obtain Mandelbrot model. Next, we generalize to the 3-dimensional case.

The 2-dimensional approximate lung model. Consider for simplicity the Euclidean space $\mathbb{R}^2$ equipped with its canonical basis (e_1, e_2), and the segment $[O, A]$, where $O = (0, 0)$ is the origin, and $A = (0, 1)$. Let next $\theta_1, \theta_2 \in (0, \frac{\pi}{2}]$ be two angles. Consider next the affine transformations T_1, and T_2 which transform the basis (e_1, e_2) to (e_1, u_1), and (e_1, v_1), respectively, where

$$u_1 \sin \theta_1 e_1 - \cos \theta_1 e_2 \text{ and } v_1 = -\sin \theta_2 e_1 - \cos \theta_2 e_2.$$

The transformations T_1, and T_2 transform the segment $[o, A]$ to the segments $[o, A_1]$, and $[o, A_2]$, respectively, as in Figure 9.4.

We notice here that both transformations T_1, and T_2 are similarities. To get contractive transformations we just multiply with ratios r_1, and r_2 in $(0, 1)$. Moreover, by choosing $\theta_1 = \theta_2$, we get a symmetric model as in Mandelbrot model illustrated in Figure 9.5.

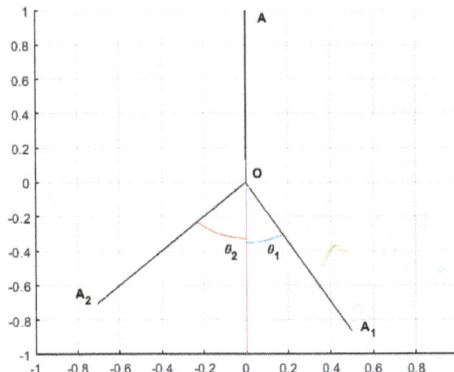

Fig. 9.4: First step for 2-dimensional Mandelbrot model.

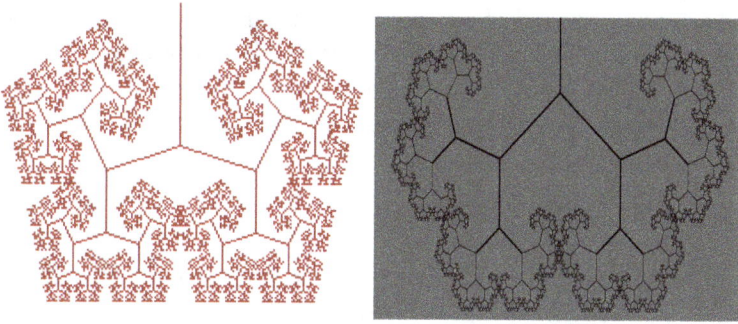

Fig. 9.5: The tree structure of human lung.

The following algorithm is a simple Matlab code to get Figure 9.4. Readers may generalize it easily to get higher order iterations, and 3-dimensional case with necessary modifications.

Algorithm 2: Mandelbrot 2-dimensional lung tree model.

Input: θ_1,θ_2,r;
Output: $[O,A],T_1([O, A)),T_2([O, A))$;
1   $R_1 = [1 \ \sin(\theta_1); 0 - \cos(\theta_1)]$;
2   $R_2 = [1 - \sin(\theta_2); 0 - \cos(\theta_2)]$;
3   $A_1 = R_1 * A$;
4   $A_2 = R_2 * A$;
5   line(OA,'color','k','Linewidth',1);
6   line(OA_1,'color','k','Linewidth',1);
7   line(OA_2,'color','k','Linewidth',1);
8   Arc = @(radius,angle) [radius*cosd(angle);radius*sind(angle)];
9   angle1 = linspace(0,θ_1);
10   radius = r;
11   Arc1 = Arc(radius,angle1);
12   plot(Arc1(1,:),Arc1(2,:));
13   angle2 = linspace(0,θ_2);
14   Arc2 = Arc(radius,angle2);
15   plot(Arc2(1,:),Arc2(2,:)).

In general, fractal geometry applications in bio-science are enormous, and also fascinating. This is due to the Capacity of fractal geometry, and/or fractal modeling in describing many organs, and phenomena in an optimal way that accurately reflect the real case. In human (and animal) bodies, fractal geometry permitted to explain, and to describe well the phenomenon of enclosing a maximal surface in a reduced volume, such as, the lung studied above, the kidneys, intestines, and also blood vessels.

9.4   Fractals for time series

This section is twofold. firstly, we aim to develop a functional, and/or statistical example to show the relationship of fractal/multifractal analysis/geometry of sets, and measures to the functional case, as raised in the previous chapters. Besides, we want to develop an application of fractal analysis to the study of time series, and how to extract their properties, and thus to show that such series may hide a fractal/multifractal structure, which is in fact, and generally related, to the case study, i.e., the real phenomenon represented by the series.

Time series may be issued from many fields, and applications, such as, finance, and economics, when modeling market indices, prices, etc, from biological applications, such as, the conversion of DNA, and protein series into numerical ones to be investigated by the mathematical tools, ECG, EEG, heart beats are also examples of such series. The presence of the fractal behaviour, and/or structure in these series is nowadays a very well-known, and confirmed concept, especially, with the discovery of fractal software, interfaces, applied in the field such the famous Fraclab software introduced originally by Vehel, and his collaborators ([Vehel and Legrand (2004)]). Figure 9.6 below is an example illustrating a financial index, where the scaling, and/or fractal law appears.

Another example is illustrated by Figure 9.7, which represents a time series modeling heart beats records. Time series may be also issued from signals, such as, vocal ones, and may also be generalized to the 2-dimensional case by representing images which may be seen as 2-dimensional time series, and or 2-dimensional signals.

Time series are also applied to forecast future situations in many phenomena, such as, markets, climate, ... In this case, the modelers applied the past of the series to forecast its future by applying mathematical expressions on the form

$$x_t = f(x_{t-1}, x_{t-2}, ..., x_{t-r}),$$

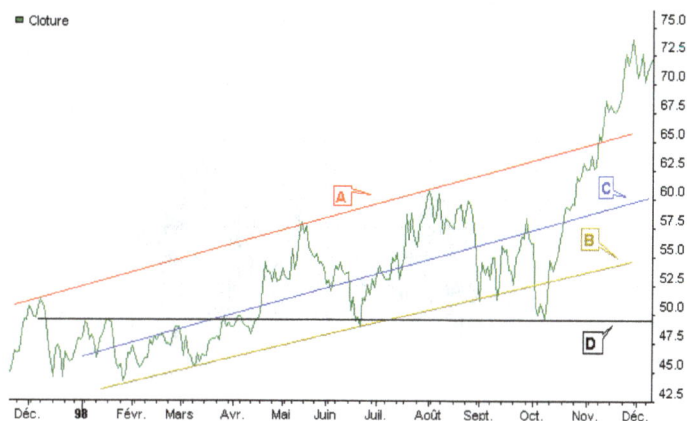

Fig. 9.6: Stock index.

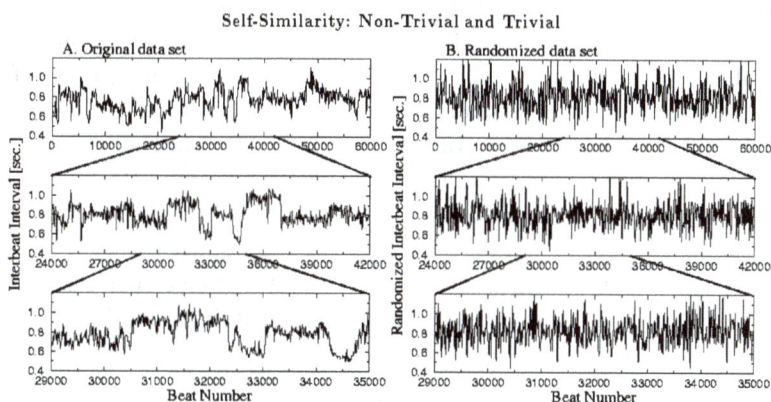

Fig. 9.7: Heart beat records.

for some parameter r fixed. The problem becomes to find the best approximator or function f satisfying the equation, and describing the best (relatively to some error measure) the reality.

In the present part, we will investigate an artificial example, which may be related to any of the cases raised above, and also to stochastic cases, such as, Brownian motion. We will consider a time series generated from a simple case of nonlinear autoregressive model, for example, on the form

$$x_{t+1} = ax_t^3 + bx_{t-1} + \epsilon_t, \tag{9.1}$$

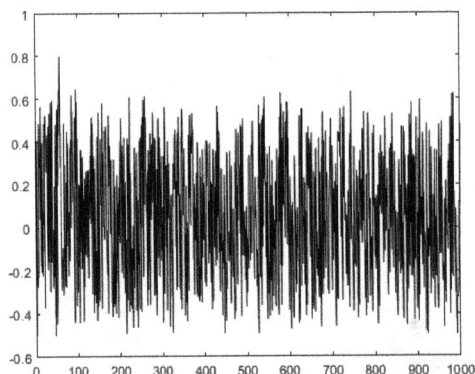

Fig. 9.8: The series $x(t)$ due to (9.1).

where a, b, c are real number parameters, and ϵ_t is a noise. Figure 9.8 is one illustration of the last model (9.1) with suitable parameters. An original way to study the fractality/multifractality of the series is to consider some special type of moments, and to extract the eventual scaling law(s) from these moments. Next, the task will be to check if certain scaling law(s) is (are) inherited by the expectancy, variance, ...etc. Such inheritance will permit to conclude on the fractal/multifractal structure of the series.

The first step consists in evaluating the so-called increments of the series computed relatively to suitable time steps. Denote for instance $I_T = [0, T]$ to be the time interval of a series $x(t)$, $t \in I_T$, and ν be a time step, which permits to subdivide the whole interval I_T into sub-intervals $I_i = [t_i, t_{i+1}]$, where

$$t_i = \nu i, \ \ 0 \leq i \leq N_\nu = \frac{T}{\nu},$$

being the number of points t_i of the subdivision. The associated increments of the series $x(t)$ are defined by

$$\Delta x_i = x(t_{i+1}) - x(t_i).$$

This gives raise to the so-called partition function, which is in fact composed by the q-increments of the series, let

$$Z(q) = \sum_{i=0}^{N_\nu} |\Delta x_i|^q.$$

The logarithmic partition function $\log Z(q)$ is next regressed against $\log \nu$ to yield the so-called scaling function $\tau(q)$, which satisfies

$$Z(q) \approx \nu^{\tau(q)}.$$

The estimation of the scaling function $\tau(q)$ yields an estimation of the multifractal spectrum

$$d(\alpha) = \inf_q(\alpha q - \tau(q)).$$

Mathematically speaking, the multifractal spectrum is based on the estimation of the regularity character of the time series. The original definition passes through the so-called Hölder regularity, which consists for a time series $x(t)$ in finding near a point t_0, the best exponent H, such that,

$$|x(t) - x(t_0)| \sim |t - t_0|^H,$$

near t_0. The Hölder regularity of $x(t)$ at the point t_0 is the maximum of all exponents H satisfying the estimation. The idea next is based on the subdivision of the whole support of the time series into sub-cells composed of the level sets according to the Hölder regularity,

$$L_h = \{t;\ H_x(t) = h\},$$

for suitable levels h.

The multifractal formalism states that the multifractal spectrum may be evaluated by means of a Legendre transform of a suitable function issued from the time series itself, such as, the partition function, and permits thus to compute the Hausdorff dimension of the level sets L_h, such as,

$$d(h) = \inf_q(hq - Z(q)).$$

Backgrounds about the multifractal spectrum may be found in [Frisch and Parisi (1985); Benzi et al (1984)].

Figure 9.9 illustrates the scaling function $\tau(q)$ associated to the series issued from equation (9.1). Figure 9.10 illustrates the multifractal spectrum of the series $x(t)$ issued from equation (9.1) and shown in Figure (9.8).

9.5 Fractals for signals/images: The case of nano images

As is known, there are many variants of fractal dimensions to be used in image and signal processing, such as image segmentation relative to the fractal dimension. The basic idea reposes on the understanding of fractal dimensions as a statistical measure, which permits to discover, and/or to check the fractal nature of objects. Nowadays, fractal dimensions are widely applied in image processing, especially, in materials-issued images in order to characterize for example the breaking up zones in materials.

Indeed, different behaviours observed in nanomaterials involve, and/or necessitate fractal models to be explored, and exploited, such as the

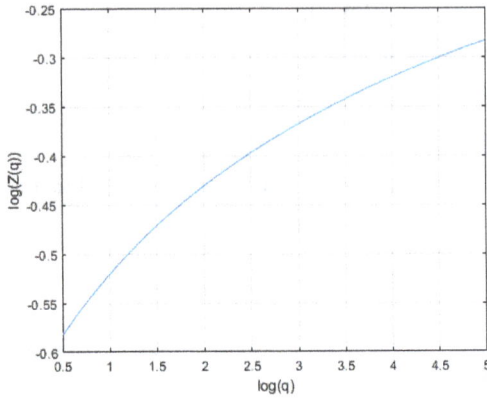

Fig. 9.9: The scaling function of the series $x(t)$ due to (9.1).

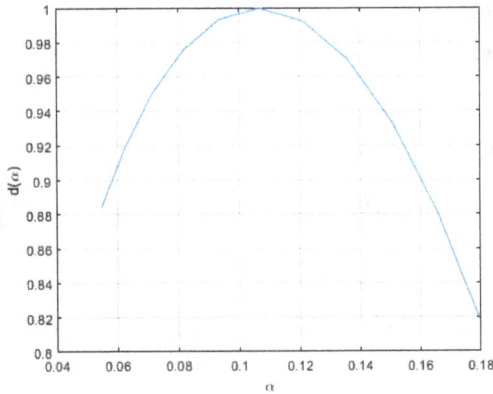

Fig. 9.10: The multifractal spectrum of the series $x(t)$ due to (9.1).

porous structure, and oscillating behaviour, which are known phenomena in materials.

Fractal analysis/geometry constitutes an excellent mathematical tool also in detecting irregularities in images. These irregularities may be due to different causes, such as, experimental conditions in laboratories, natural effects, etc.

Mathematical investigations of the behavior of materials has been widely studied for classical cases, and composites. However, the use of modern and sophisticated mathematical tools such as, fractals, wavelets, fractional calculus in nanomaterials, are somehow recent, and has been included in such topics at the last few decades. See [Baleanu (2012a,b, 2015); Baleanu et al (2010); Cattani and Karaca (2018); Cattani and Rushchitsky (2007)]. In parallel, experimental studies, however, have been growing up rapidly, even for nanomaterials, compared to their theoretical explanations by mathematical, and theoretical physics tools. Nanomedicine, for example, is nowadays an attractive field for researchers from both theory and applications. There are for example essays of implementing plasmonic materials' nanoparticles for cancer diagnosis. Such implementation may induce noisy signals. Scientists search to apply noble materials that may be most noiseless, such as, gold ([Huang and El-Sayed (2010)]). In giant magneto-resistance, the oscillating behaviour of multilayer nanofilms has been noticed (See [Zhang (2018)]). Recent investigations have been also done for fluidization of nanoparticles in the presence of an oscillating magnetic field [Yu et al (2005)]. It is shown that the bed of nanoparticle agglomerates may be smoothly fluidized, and that the minimum fluidization velocity may be significantly reduced relatively to frequencies of the oscillating magnetic field.

Now, with the technological development, and implementation of modern instruments, such as, the nano microscopes, the physical properties of materials becomes more and more comprehensive. Moreover, new materials, and new properties have been discovered. The fractal structure is one of the physical properties expressed at the super molecular level, and which is more explored by nano microscopes.

In the present part, we propose to conduct some discussions about the use of fractal dimensions for nanomaterials' images. This will permit to reach at the same time, the use of fractals in both image/signals, and nanomaterials' applications, we propose to describe how fractals are used in understanding nanoimages; those images issued from nanomaterials.

Estimating the fractal dimension of images will permit to describe well the structural properties. The most (and even simple) used method to estimate the fractal dimension of an image is the so-called Box-dimension, known also the Bouligand–Minkowski dimension. It is based on a simple mathematical formula that computes the log-log slope of the maximum number of boxes (squares, balls, cubes, circles, ...) used to cover the image (black pixels for example, when the image is converted to black and white), by the size of such boxes. The Figure 9.11 illustrates the principle of counting the box-dimension of an image (see https://en.wikipedia.org/wiki/Minkowski-Bouligand-dimension), and in Figure 9.12, (see [Brown et al (2005)]).

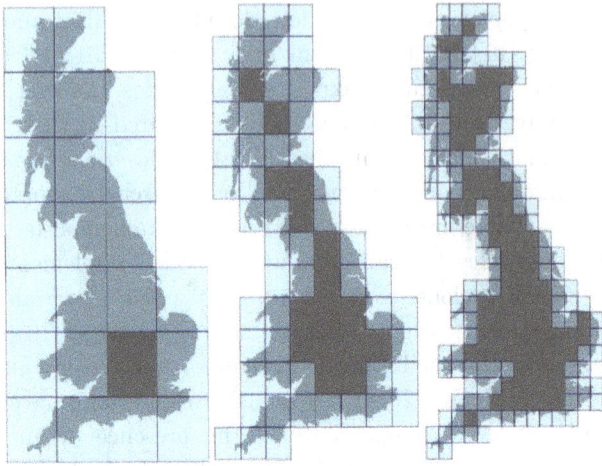

Fig. 9.11: The box dimension estimation of the Great Britain coast.

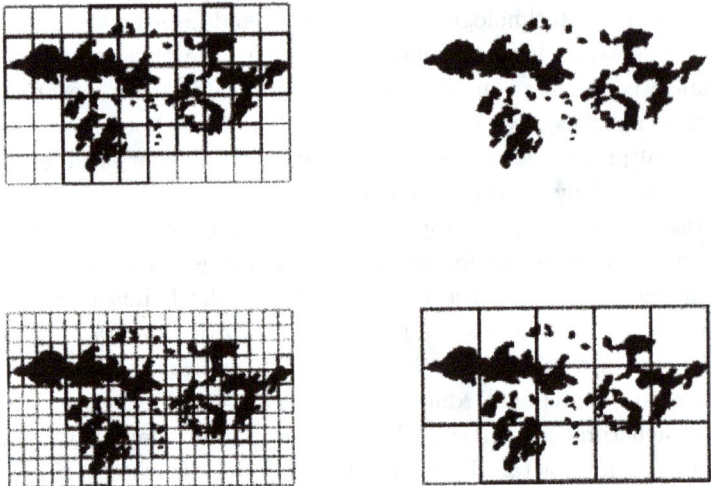

Fig. 9.12: The box dimension estimation of a black-white image.

In the present section, we consider the following image presented in Figure 9.13, and which designates an image of Titanium dioxide TiO_2 nanoparticles. Such nanomaterials are well known, and have many important properties. In industrial aims, they are capable to add new functionalities to infrastructures, such as self-cleaning properties. They serves also

Fig. 9.13: Titanium dioxide TiO_2 nanoparticles.

Table 9.1:   Estimation of the fractal dimension of TiO_2
illustrated in Figure 9.13.

Scale	Iteration	Fitting coef	dim
2	5	0.0534	1.9615
2	6	0.0615	1.9594
2	9	0.0583	1.9689
3	5	0.0734	1.9623
3	6	0.0639	1.9674
3	9	0.0323	1.9778

to remove air pollutants through photocatalysis. In building constructions, TiO_2 are used to degrade organic pollutants, without affecting the aesthetic characteristics of concrete structures. Titanium dioxide nanoparticles are particles of TiO_2, with small diameters less than 100 nm. In health aims, these nanoparticles are applied, for example, in ultrafine form. Indeed, TiO_2 are used in sunscreens due to its ability to block UV radiation while remaining transparent on the skin. The health risks of ultrafine TiO2 from dermal exposure on intact skin are considered extremely low, and it is considered safer than other substances used for UV protection.

In Table 9.1, estimations of the fractal dimension due to box-counting method are provided. A log-log model relating the number of boxes to cover the image, denoted here by N_s to the size of boxes is applied on the form

$$\log(N_s) = -a\log(Size) + b,$$

where the size is computed by means of dyadic cubes 2^{-s}, and triadic 3^{-s}, and b is a fitting coefficient, assuring the regression. The coefficient (slope) a

will be eventually the estimated dimension. We notice easily from Table 9.1 a mean value of the dimension estimated as

$$dim \approx 1.9646.$$

9.6   A classical fractal self-similar set

The most simple, comprehensible, and original example in fractal sets may be the example of the triadic Cantor set. For this reason, we will consider a self-similar analogue of this set in dimension 2, and show how we can apply the theories presented in the previous chapters to deduce the characteristics of the Cantor-type set obtained. The example will be about the well known Sierpinski carpet due to [Sierpinski (1916)]. One of the important remarks will be about a comparison between the fractal dimension of the 2-dimensional Sierpinski carpet, and the one of the Cartesian product of Cantor sets.

The Sierpinski carpet is a plane fractal described obtained as follows. Consider the unit square $I_0 = [0, 1] \times [0, 1]$. In a first step, we by subdivide it into nine isometric squares, and remove the middle one. Next, we subdivide each one of the eight remaining squares into nine isometric squares, and remove again the new middle squares. We next continue recursively the same process. The resulting limit set ($n \to \infty$, where n is the number of steps or iterations) is the so-called Sierpinski carpet. Figure 9.14 illustrates the process.

Mathematically speaking, the Sierpinski carpet may be deduced as the invariant set associated to the iterated function system due to the following eight similarities defined on $\mathbb{R}^2$ by

$$S_1 \begin{pmatrix} x \\ y \end{pmatrix} = \begin{pmatrix} 1/3 & 0 \\ 0 & 1/3 \end{pmatrix} \begin{pmatrix} x \\ y \end{pmatrix}$$

$$S_2 \begin{pmatrix} x \\ y \end{pmatrix} = \begin{pmatrix} 1/3 & 0 \\ 0 & 1/3 \end{pmatrix} \begin{pmatrix} x \\ y \end{pmatrix} + \begin{pmatrix} 0 \\ 1/3 \end{pmatrix}$$

$$S_3 \begin{pmatrix} x \\ y \end{pmatrix} = \begin{pmatrix} 1/3 & 0 \\ 0 & 1/3 \end{pmatrix} \begin{pmatrix} x \\ y \end{pmatrix} + \begin{pmatrix} 0 \\ 2/3 \end{pmatrix}$$

$$S_4 \begin{pmatrix} x \\ y \end{pmatrix} = \begin{pmatrix} 1/3 & 0 \\ 0 & 1/3 \end{pmatrix} \begin{pmatrix} x \\ y \end{pmatrix} + \begin{pmatrix} 1/3 \\ 0 \end{pmatrix}$$

$$S_5 \begin{pmatrix} x \\ y \end{pmatrix} = \begin{pmatrix} 1/3 & 0 \\ 0 & 1/3 \end{pmatrix} \begin{pmatrix} x \\ y \end{pmatrix} + \begin{pmatrix} 1/3 \\ 2/3 \end{pmatrix}$$

Fig. 9.14: The first 6 steps of the construction of Sierpinski carpet.

$$S_6 \begin{pmatrix} x \\ y \end{pmatrix} = \begin{pmatrix} 1/3 & 0 \\ 0 & 1/3 \end{pmatrix} \begin{pmatrix} x \\ y \end{pmatrix} + \begin{pmatrix} 2/3 \\ 0 \end{pmatrix}$$

$$S_7 \begin{pmatrix} x \\ y \end{pmatrix} = \begin{pmatrix} 1/3 & 0 \\ 0 & 1/3 \end{pmatrix} \begin{pmatrix} x \\ y \end{pmatrix} + \begin{pmatrix} 2/3 \\ 1/3 \end{pmatrix}$$

$$S_8 \begin{pmatrix} x \\ y \end{pmatrix} = \begin{pmatrix} 1/3 & 0 \\ 0 & 1/3 \end{pmatrix} \begin{pmatrix} x \\ y \end{pmatrix} + \begin{pmatrix} 2/3 \\ 2/3 \end{pmatrix}$$

acted on the square I_0. Denote $\mathbb{S}$ the Sierpinski carpet. We immediately have

$$\mathbb{S} = S_1(\mathbb{S}) \cup S_2(\mathbb{S}) \cup S_3(\mathbb{S}) \cup S_4(\mathbb{S}) \cup S_5(\mathbb{S}) \cup S_6(\mathbb{S}) \cup S_7(\mathbb{S}) \cup S_8(\mathbb{S}).$$

The Sierpinski carpet can be realized as the set of points in the unit square whose coordinates written in base three do not both have a digit '1' in the same position, using the infinitesimal number representation.

Remark easily that for $i = 1, ..., 8$, we have

$$\left\| S_i \begin{pmatrix} x \\ y \end{pmatrix} - S_i \begin{pmatrix} x' \\ y' \end{pmatrix} \right\| = \frac{1}{3} \left\| \begin{pmatrix} x \\ y \end{pmatrix} - \begin{pmatrix} x' \\ y' \end{pmatrix} \right\|,$$

which means that the S_i's, $i = 1, ..., 8$ are similarities with (the same) ratios

$$r_1 = r_2 + ... = r_8 = \frac{1}{3}.$$

As a consequence, by applying the theory of fractal dimensions of self-similar sets, we deduce that all the dimensions of the Sierpinski carpet (Hausdorff, packing, and Box dimension) coincide, and are equal to

$$dim\,\mathbb{S} = s,$$

where s is the unique real number satisfying

$$r_1^s + r_2^s + ... + r_8^s = 1,$$

which is equivalent to

$$dim\,\mathbb{S} = s = \frac{\log 8}{\log 3}.$$

Moreover, the Sierpinski carpet is a compact subset of the plane, with Lebesgue measure 0, with its interior being empty. A simple, and good exercise may be to develop a simple algorithm yielding the Sierpinski carpet described above. Besides, different forms may be obtained by applying different linear parts for the similarities above, such as, homotheties composed with rotations, and/or random variants.

9.7 A case of self-similar type measures

One of the important, and widely studied concept in fractal, and multi-fractal analysis is the notion of self-similar measures. These measures are strongly related to self-similar sets, such as, the Cantor, and Sierpinski, which have been exposed previously. We will develop in this section a simple example of self-similar measures, and show their properties, such as, the computation of the fractal dimension. Indeed, let $S = (S_i)_{1 \le i \le m}$ be m similarities defined on $[0, 1]$, ($m \in \mathbb{N}$ fixed). Let also $p = (p_1, p_2, ..., p_m)$ be a probability vector ($p_i \ge 0$, such that, $p_1 + p_2 + ... + p_m = 1$). We say that a measure μ on $\mathbb{R}$ is self-similar associated to the system (S, p) if it satisfies the equation

$$\mu = \sum_{i=1}^{m} p_i \mu \circ S_i^{-1}. \tag{9.2}$$

In fact the measure μ is unique. Moreover, when the vector p is strictly positive, the measure μ is topologically supported by the unique self-similar set (Cantor set) associated to the similarity set S, which is also the unique non-empty compact invariant set relative to S. Denote $\mathbb{S}_\mu = \text{support}(\mu)$. It satisfies uniquely

$$\mathbb{S}_\mu = S_1(\mathbb{S}_\mu) \cup S_2(\mathbb{S}_\mu) \cup ... \cup S_m(\mathbb{S}_\mu).$$

The multifractal analysis of the measure μ passes through the evaluation of its spectrum of singularities. The first step consists in partitioning its support into level sets

$$S(\alpha) = \left\{ x \in \mathbb{S}_\mu; \lim_{r \downarrow 0} \frac{\log(\mu(B(x, r)))}{\log r} = \alpha \right\}, \ \alpha \geq 0.$$

Whenever the contractions or similitudes S_i satisfies some condition known as the open set condition, the computation of the spectrum of singularities is possible, and exact. Denote, for $t \in \mathbb{R}$, $\beta(t)$ the unique real number satisfying

$$\sum_{i=1}^{m} p_i^t r_i^{\beta(t)} = 1,$$

where the r_i's, $i = 1, ..., m$, are the ratios of the contractive similitudes S_i, $i = 1, ..., m$, respectively, and assume the open set condition

$$S_i(I) \subset]0, 1[, \forall i, \quad \text{and} \quad S_i(I) \cap S_j(I) = \emptyset, \forall i, j; \ i \neq j.$$

The measure μ consists in affecting each interval $I_i = S_i(I)$ with the weight p_i. The function $\beta(t)$ is differentiable, and

$$\beta'(t) = -\frac{\displaystyle\sum_{i=1}^{m} p_i^t r_i^{\beta(t)} \log p_i}{\displaystyle\sum_{i=1}^{m} p_i^t r_i^{\beta(t)} \log r_i}.$$

This permits the computation next of the exact spectrum of singularities of μ. We will apply here the result of Chapter 7.

For $n \in \mathbb{N}$, and $i = (i_1, i_2, ..., i_n) \in \{1, \ 2, \ ..., m\}^n$, denote

$$S_i = S_{i_n} \circ S_{i_{n-1}} \circ .... \circ S_{i_1} , \ r_i = r_{i_1} r_{i_2} .... r_{i_n} , \ p_i = p_{i_1} p_{i_2} .... p_{i_n}.$$

Denote G_n the set of intervals $I_i = S_i(I)$ (called also nth generation). Consider next the values $m = 7$,

$$p_1 = p_4 = p_7 = \frac{1}{3}, \ p_2 = p_3 = p_5 = p_6 = 0 \text{ and } r_i = \frac{1}{7}, \ \forall i.$$

It is easy to check that

$$\beta(t) = (1 - t)\frac{\log 3}{\log 7}.$$

Let ν be the measure defined by

$$\nu(I_i) = p_i^t r_i^{\beta(t)} \quad \text{and} \quad 0 \text{ elsewhere.}$$

It consists of a probability measure on I with the same support as μ. Let $x \in \mathbb{S}_\mu$, $r > 0$, and $n \in \mathbb{N}$ be, such that,

$$\frac{1}{7^n} \le 2r < \frac{2}{7^n}.$$

There exists a multi-index $i = (i_1, i_2, ..., i_n) \in \{1, 4, 7\}^n$, such that, I_i is the unique interval of G_n, which intersects $I(x, r)$. Such interval has surely a predecessor I_j of G_{n+1} contained in $I(x, r)$. ($j \in \{1, 4, 7\}$). It holds consequently that

$$\mu(I_{ij}) \le \mu(I(x, r)) \le \mu(I_{ij}),$$

and

$$\nu(I_{ij}) \le \nu(I(x, r)) \le \nu(I_{ij}).$$

Denote next

$$\underline{K} = \inf \left\{ (\min_i p_i)^t (\min_i r_i)^{\beta(t)}, \ \frac{(\min_i p_i)^t (\max_i r_i)^{\beta(t)}}{(\min_i r_i)^{\beta(t)}}, \ \frac{(\max_i p_i)^t (\max_i r_i)^{\beta(t)}}{(\min_i p_i)^{\beta(t)}} \right\}$$

and

$$\overline{K} = \sup \left\{ \frac{1}{(\min_i p_i)^t (\min_i r_i)^{\beta(t)}}, \ \frac{1}{(\min_i p_i)^t}, \ \frac{1}{(\min_i r_i)^{\beta(t)}} \right\}.$$

We have

$$\underline{K} \le \frac{\nu_q(I(x, r))}{\left[\mu(I(x, r))\right]^q (2r)^t} \le \overline{K}.$$

Using Theorem 8.3, we obtain

$$C_q(t) = -t\frac{\log 3}{\log 7}.$$

It is a differentiable function, with,

$$C_q'(t) = -\frac{\log 3}{\log 7}.$$

From Corollary 8.2, we obtain

$$dim\, X(\frac{\log 3}{\log 7}) = \frac{\log 3}{\log 7}.$$

9.8 Exercises for Chapter 9

Exercise 1.
Consider the Sierpinski triangle $\mathcal{ST}$ constructed from the equilateral triangle of vertices $O(0,0)$, $A(1,0)$, and $B(\frac{1}{2}, \frac{\sqrt{3}}{2})$ in the plane by subdividing it into four smaller congruent equilateral triangles, and removing the central triangle, and repeating the same removal action with each of the remaining smaller triangles infinitely.
(1) Find three similarities on $\mathbb{R}^2$ yielding $\mathcal{ST}$ as the associated unique non-empty compact invariant set.
(2) Write an algorithm (code) to illustrate the process graphically.
(3) Compute the Hausdorff dimension of $\mathcal{ST}$.

Exercise 2.
Denote $\mathcal{K}$ the Koch snowflake constructed from the equilateral triangle ABC in the plane, with vertices $O(0,0)$, $A(1,0)$, and $B(\frac{1}{2}, \frac{\sqrt{3}}{2})$ by dividing each line segment into three segments of equal length to draw an equilateral triangle with the middle segment as its base, and points outward, and removing the line segments that are bases of the new triangles. The Koch snowflake $\mathcal{K}$ is the limit as the above steps are followed infinitely.
(1) Find suitable similarities on $\mathbb{R}^2$ yielding $\mathcal{K}$ as the associated unique non-empty compact invariant set.
(2) Write an algorithm (code) to illustrate the process graphically.
(3) Compute the perimeter of $\mathcal{K}$.
(4) Compute the area of $\mathcal{K}$.
(5) Compute the Hausdorff dimension of $\mathcal{K}$.
(6) Compute the volume of the solid of revolution of $\mathcal{K}$ about the axis of symmetry $x = \frac{1}{2}$ of the triangle OAB.

Exercise 3.
Let $f : [0,1] \to \mathbb{R}$ be, such that,

$$f(x) = 3x\chi_{[0,1/2[}(x) + 3(1-x)\chi_{[1/2,1]}(x).$$

For $n \in \mathbb{N}$, denote $f^n = F \circ f \circ ... \circ F$, n times, and let

$$\Lambda_n = \{x \in [0,1]; \ f^k(x) \in [0,1], \ \text{for } k = 0, 1, ..., n\},$$

and

$$\Lambda = \{x \in [0,1]; \ f^n(x) \in [0,1], \ \text{for all } n \geq 0\}.$$

(1) Draw the graphs of f, Λ_1, and Λ_2.

(2) Give the endpoints of the intervals in Λ_1, and Λ_2.

(3) Give the endpoints of the intervals in Λ_n.

(4) Is Λ a Cantor set? Justify your answer.

(5) Compute the Lebesgue measure of Λ.

(6) Compute the box counting dimension of Λ.

(8) Find an iterated function system whose limit is Λ.

Exercise 4.

Consider in $\mathbb{R}^2$ the set X constructed as follows. We start by the square $[0,1] \times [0,1]$. Next, in the first iteration, we subdivide the original square into nine equal squares, and remove the upper right two squares, and the bottom left two squares (in the same direction, horizontal or vertical both of them). In the second iteration, we subdivide each of the remaining squares into nine equal squares, and remove the upper right two squares, and the bottom left two squares by the same way as in iteration 1. We continue to infinity.

(1) Compute the Lebesgue measure of X.

(2) Compute the box counting dimension of X.

(3) Find an iterated function system whose limit is X.

Exercise 5.

Consider the so-called fractal antenna $\mathcal{FA}$, applied in cellular phones, and constructed as illustrated by a deterministic iterated function system program, with the five first steps of iteration as in Figure 9.15 below.

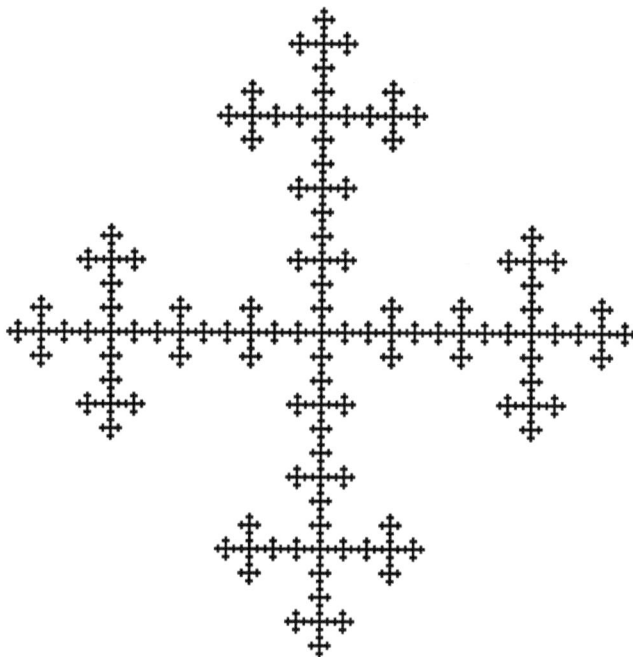

Fig. 9.15: The first five iterations of a fractal antenna.

(1) Find the smallest number of transformations needed for this iterated function system.

(2) Write the iterated function system rules.

(3) Compute the Lebesgue measure of $\mathcal{FA}$.

(4) Compute the box counting dimension of $\mathcal{FA}$.

Bibliography

Alimohammady, M., Cattani, C. and Kalleji, M. K. (2017) Invariance and existence analysis for semilinear hyperbolic equations with damping and conical singularity. J. Math. Anal. Appl., 455(1), 569–591.

Anguera, J., Andujar, A., Jayasinghe, J., Chakravarthy, V. and Chowdary, P. S. R. (2020) Fractal Antennas: An Historical Perspective. Fractal and Fract., 4(1).

Arfaoui, S., Rezgui, I. and Ben Mabrouk, A. (2017) Wavelet Analysis On The Sphere, Spheroidal Wavelets, Degryuter, (with Sabrine Arfaoui and Imen Rezgui). Degruyter, 2017, ISBN 978-3-11-048188-4.

Aversa, V. and Bandt, C. (1990) The Multifractal Spectrum of Discrete Measures. Acta Univ. Carolinae-Math. Et. Phys., 31(2), 5–8.

Azizieh, C. (2002) Modélisation de Séries Financières par un Modèl Multifractal. Mémoire pour diplôme d'actuaire. Université Libre de Bruxelles.

Badea, A. F., Platon, M. L., Crisan, M., Cattani, C., Badea, I. and Pierro, G. (2013) Fractal analysis of elastographic images for automatic detection of diffuse diseases of salivary glands: Preliminary results. Computational and Mathematical Methods in Medicine, 1–6.

Baker, A. and Schmidt, W. M. (1970) Diophantine approximation and Hausdorff dimension. Proc. London Math. Soc., 21, 1–11.

Baker, R. C. (1976) Dyadic Methods in the measure theory of numbers. Transactions of the AMS, 221, 419–432.

Baleanu, D. (2012) Wavelet Transforms and Their Recent Applications in Biology and Geoscience. InTechOpen, 2012.

Baleanu, D. (2012) Advances in Wavelet Theory and Their Applications in Engineering, Physics and Technology. InTechOpen, 2012.

Baleanu, D. (2015) Wavelet Transform and Some of Its Real-World Applications. InTechOpen, 2015.

Baleanu, D., Guvenc, Z. B. and Tenreiro Machado, J. A. (2010) New Trends in Nanotechnology and Fractional Calculus Applications. Springer.

Baleanu, D., Machado, J. A. T., Cattani, C., Baleanu, M. C. and Yang, X. J. (2014) Local fractional variational iteration and decomposition methods for wave equation on Cantor sets within local fractional operators. Abstr. Appl. Anal., 82, 1–6.

Barlow, M. T. and Taylor, S. J. (1992) Defining fractal subsets of $\mathbb{Z}^d$. Proc. London Math. Soc., 64(3), 125–152.

Barnsley, M. F. (2000) Fractals Everywhere. Second edition, Morgan Kaufmann.

Barnsley, M. F., Devaney, R. L., Mandelbrot, B. B., Peitgen, H.-O., Saupe, D. and Voss, R. F. (1988) The Science of Fractal Images, Heinz-Otto Peitgen. In: Dietmar Saupe (eds.), Springer-Verlag New York.

Barral, J., Nasr, F. B. and Peyrière, J. (2003) Comparing multifractal formalism: The neighbouring box condition. Asian J. Math., 7, 149–66.

Batakis, A. and Heurteaux, Y. (2002) On relations between entropy and hausdorff dimension of measures. Asian J. Math., 6(3), 399–408.

Beardon, A. F. (1965) On the Hausdorff dimension of general Cantor sets. Proc. Camb. Phil. Soc., 61, 679–694.

Bélair, J. (1987) Sur le calcul de la dimension fractale. Ann. SC. math. Québec, 11(1), 7–23.

Benaych-Georges, F. (2009) Introduction au calcul stochastique pour la finance, Cours, Master 1, Université Paris VI, 2009.

Ben Mabrouk, A. (2005) Multifractal analysis of some non isotropic quasi-self-similar functions. Far East J. Dyn. Syst., 7(1), 23–63.

Ben Mabrouk, A. (2006) Wavelet Analysis of Anisotropic Quasi-Self-Similar Functions in a Nonlinear Case, Colloque WavE 2006, Polytechnic Federal School of Lausanne, Switzerland, 10–14 July.

Ben Mabrouk, A. (2007) Study of some nonlinear self-similar distributions. International Journal of Wavelets, Multiresolution and Information Processing, 5(6), 1–10.

Ben Mabrouk, A. (2008) Higher order multifractal formalism. Stat. Proba. Lett., 78, 1412–1421.

Ben Mabrouk, A. (2008) On some nonlinear non isotropic quasi self similar functions. Nonlinear Dyn., 51, 379–398.

Ben Mabrouk, A. (2008) An adapted group dilation anisotropic multifractal formalism for functions. J. Nonlinear Math. Phys., 15(1), 1–23.

Ben Mabrouk, A. (2008) Anouar Ben Mabrouk: Higher order multifractal measures. Stat. Probab. Lett., 78, 1412–1421.

Ben Mabrouk, A. (2008) Wavelet analysis of nonlinear self-similar distributions with oscillating singularity. Int. J. Wavelets Multiresolution Inf. Process., 6(3), 1–11.

Ben Mabrouk, A. and Ben Abdallah, N. (2006) Study of some multinomial cascades. Adv. Appl. Stat., 6(3), 95–303.

Ben Mabrouk, A., Ben Abdallah, N. and Dhifaoui, Z. (2008) Wavelet decomposition and autoregressive model for the prevision of time series. Appl. Math. Comput., 199(1), 334–340.

Ben Mabrouk, A., Ben Abdallah, N. and Hamrita, M. E. (2011) A wavelet method coupled with quasi self similar stochastic processes for time series approximation. Int. J. Wavelets, Multiresolution Inf. Process., 9(5), 685–711.

Ben Mabrouk, A. and Aouidi, J. (2011) Multifractal analysis of weighted quasi-self-similar functions. Int. J. Wavelets Multiresolution Inf. Process., 9(6), 965–987.

Ben Mabrouk, A. and Aouidi, J. (2012) Lecture Note on Wavelet Multifractal Analysis of Self Similarities. Lampert Academic Publishing, Verlag. ISBN 978-3-8465-9646-3.

Ben Mabrouk, A., Aouidi, J. and Ben Slimane, M. (2014) A wavelet multifractal formalism for simultaneous singularities of functions. Int. J. Wavelets, Multiresolution Inf. Process., 12(1).

Ben Mabrouk, A., Aouidi, J. and Ben Slimane, M. (2016) Mixed multifractal analysis for functions: General upper bound and optimal results for vectors of self-similar or quasi-self-similar of functions and their superpositions. Fractals, 24(4), 1650039.

Ben Nasr, F. (1994) Analyse multifractale de mesures, CRAS. Paris, 319(I), 807–810.

Ben Nasr, F. (1997) On some regularity conditions of Borel measures on $\mathbb{R}$. J. Austral. Math. soc. Série A, 63, 218–224.

Ben Nasr, F. and Bhouri, I. (1997) Spectre multifractal de mesures boréliènnes sur $\mathbb{R}^d$. C. R. Acad. Sci. Paris, t. 325, Série I, 253–256.

Ben Nasr, F., Bhouri, I. and Heurteaux, Y. (2002) The validity of the multifractal formalism: Results and examples. Adv. Math., 165, 264.

Benzi, R., Paladin, G., Parisi, G. and Vulpiani, A. (1984) On the multifractal nature of fully developed turbulence and chaotic systems. J. Phys, A-17, 3521–3531.

Berglund, N. (2014) Martingales et calcul stochastique, Master 2 Recherche de Mathéematiques, Université d'Orléans.

Berglund, N. and Gentz, B. (2006) Noise-Induced Phenomena in Slow-Fast Dynamical Systems A Sample-Paths Approach, Probability and Its Applications. Springer.

Bernay, M. (1975) La dimension de Hausdorff de l'ensemble des nombres r-déterministes. C.R. Acad. Sc. Paris, Série A, 280, 539–541.

Besicovitch, A. S. (1934) On the sum of digits of real numbers represented in the dyadic system. Math. Ann., 110, 321–330.

Beyer, W. A. (1962) Hausdorff dimension of level sets of some redemacher series. Pacific J., 12, 35–46.

Billingsley, P. (1960) Hausdorff dimension in probability theory. III. J. Math., 4, 187–209.

Billingsley, P. (1961) Hausdorff dimension in probability theory. II. III. J. Math., 5, 291–298.

Billingsley, P. (1965) Ergodic Theory and Information. Wiley, NewYork.

Billingsley, P. (1968) Convergence of Probability Measures. Wiley, NewYork.

Billingsley, P. and Henningsen, I. (1975) Hausdorff dimension of some continued-fraction sets. Z. Wahrscheinlichkeitstheorie verw. Geb., 31, 163–173.

Biswas, A., Cresswell, H. P. and Bing, C. S. (2012) Application of Multifractal and Joint Multifractal Analysis in Examining Soil Spatial Variation: A Review. Chapter 6 in Fractal Analysis and Chaos in Geosciences. Edited by Sid-Ali Ouadfeul, InTechOpen, pp. 109–138. DOI.org/10.5772/51437.

Boyd, D. W. (1973) The residual set dimension of the Apollonian packing. Mathematika, 20, 170–174.

Bozkus, S. K., Kahyaoglu, H. and Lawali, A. M. M. (2020) Multifractal analysis of atmospheric carbon emissions and OECD industrial production index. International Journal of Climate Change Strategies and Management, 12(4), 411–430. DOI: 10.1108/IJCCSM-08-2019-0050.

Briggs, J. P. (1992) Fractals: The Patterns of Chaos: Discovering a New Aesthetic of Art, Science, and Nature. Simon & Schuster.

Brown, C., Witschey, W. R. T. and Liebovitch L. S. (2005) The broken past: Fractals in archaeology. J. Archaeol. Method Th., 12(1), 37–78.

Brown, G., Michon, G. and Peyrièrre, J. (1992) On the multifractal analysis of measures. J. Stat. Phys., 66, 775–779.

Buck, R. J. (1970) A generalized Hausdorff dimension for functions and sets. Pacific J., 33, 170–174.

Buck, R. J. (1973) Hausdorff dimensions for compact sets in $\mathbb{R}^n$. Pacific J., 44, 421–434.

Budzinski, T. (2019) Processus aléatoires, TD. ENS Paris, 2017–2019.

Cabrelli, C. A., Hare, K. E. and Molter, U. M. (1997) Sums of cantor sets. Ergod. Theory Dyn. Syst., 17(6), 1299–1313.

Cabrelli, C. A., Hare, K. E. and Molter, U. M. (2002) Sums of Cantor Sets yielding an interval. J. Aust. Math. Soc. Series A, 73(3), 405–418.

Cabrelli, C. A., Mendevil, F., Molter, U. M. and Shonkwiler, R. (2004) On the Hausdorff h-Measure of Cantor Sets. Pac. J. Math., 217(1), 45–59.

Cabrelli, C. A., Molter, U. M., Paulauskas, V. and Shonkwiler, R. (2005) The Hausdorff dimension of p-Cantor sets. Real Analysis Exchange, 30(2), 413–433.

Cajar, H. (1981) Billingsley dimension in probability spaces. Lecture notes in mathematics, 892. New York: Springer.

Calvet, L. and Fisher, A. (2008) Multifractal Volatility, Theory, Forecasting, and Pricing. Academic press advanced finance series, 1st edn., September.

Castiglioni, P. and Faini, A. (2019) A fast DFA Algorithm for multifractal multiscale analysis of physiological time series. Front. Physiol., 10, 18, Article: 115. DOI: 10.3389/fphys.2019.00115.

Cattani, C. and Karaca, Y. (2018) Computational Methods for Data Analysis. De Gruyter.

Cattani, C. and Rushchitsky, J. (2007) Wavelet and wave analysis as applied to materials with micro or nanostructure. Series on Adv. Math. Appl. Sci., 74.

Cattani, C. and Pierro, G. (2013) On the fractal geometry of DNA by the binary image analysis. Bull. Math. Biol., 75(9), 1544–1570.

Cattani, C. and Ciancio, A. (2016) On the fractal distribution of primes and prime-indexed primes by the binary image analysis. Physica A: Statistical Mechanics and its Applications, 460, 222–229.

Cattani, C., Pierro, G. and Altieri, G. (2012) Entropy and multifractality for the myeloma multiple TET 2 gene. Mathematical Problems in Engineering.

Cattani, C. (2010) Fractal patterns in prime numbers distribution. International Conference on Computational Science and Its Applications, 164–176.

Cattani, C. (2010) Fractals and hidden symmetries in DNA. Mathematical Problems in Engineering 2010.

Cattani, C. (2020) Cantor waves for Signorini hyperelastic materials with cylindrical symmetry. Axioms, 9(1).

Cattani, C., Ehler, M., Li, M., Liao, Z. and Hooshmandasl, M. (2014) Scaling, self-similarity, and systems of fractional order. Abstract and Applied Analysis.

Cattani, C. and Laserra, E. (2010) Self-similar hierarchical regular lattices. International Conference on Computational Science and Its Applications, 225–240.

Cattani, C., Laserra, E. and Bochicchio, I. (2012) Simplicial approach to fractal structures. Mathematical Problems in Engineering, 1–21.

Cawley, R. and Mauldin, R. D. (1992) Multifractal decomposition of Moran fractals. Adv. Math., 92, 196–236.

Chen, Z. Y., Cattani, C. and Zhong, W. P. (2014) Signal processing for nondifferentiable data defined on Cantor sets: A local fractional Fourier series approach. Advances in Mathematical Physics.

Chen, G. and Cheng, Q. (2017) Fractal density modeling of crustal heterogeneity from the KTB deep hole. J. Geophys. Res. Solid Earth, 122, 1919–1933. Doi:10.1002/2016JB013684.

Choquet, G. (1969) Lectures on Analysis. Vol II. Benjamin. New York.

Chu, P. C. (1999) Multifractal analysis of the southwestern iceland sea surface mixed layer thermal structure. Thirteenth Symposium on Boundary Layers and Turbulence, American Meteorological Society, 476–479. http://hdl.handle.net/10945/36212.

Cole, J. (2000) Relative analysis. Choas Solitons Fractals, 11, 2233–2250.

Cole, J. and Olsen, O. (2003) Multifractal variation measures and multifractal density theorems. Real. Anal. Exch., 28, 501–514.

Colebrook, C. M. (1970) The Hausdorff dimension of certain sets of nonnormal numbers. Michigan Math. J., 17, 103–116.

Collet, P., Lebowitz, J. and Porzio, A. (1987) The dimension spectrum of some dynamical systems. J. Stat. Phys, T., 47, 609–644.

Dai, D. (1995) Frostman lemma for Hausdorff measure and packing measure in probability spaces. J Nanjing Univ Math Biquarterly, 12, 191–203.

Dai, C. (2008) On the equivalence of the multifractal centered Hausdorff measure and the multifractal packing measure. Nonlinearity, 21, 1443–1453.

Dai, M. and Jiang, Y. (2009) The equivalence of the multifractal on cookie-cutter-like sets. Chaos Solitons Fractals, 41, 1408–1415.

Dai, C. and Li, Y. (2006) A multifractal formalism in a probability space. Chaos Solitons Fractals, 27, 57–73.

Dai, M. and Liu, Z. (2008) The quantization dimension and other dimensions of probability measures. Int. J. Nonlinear Sci., 5, 267–274.

Das, M. (1998) Local properties of self-similar measures. Illinois J. Math., 42, 313–332.

Das, M. (2005) Hausdorff measures, dimensions and mutual singularity. Trans. Am. Math. Soc., 357, 4249–4268.

David, G. and Semmes, S. (1997) Fractured Fractals and Broken Dreams. Clarendon Press, Oxford.

Debussche, A. (1998) Hausdorff dimension of random invariant set. JPAM., 77, 967–988.

Dolotin, V. and Morozov, A. (2006) Universal Mandelbrot Set: Beginning of the Story, World Scientific Publishing Company.

Dryakhlov, A. V. and Tempelman, A. A. (2001) On Hausdorff dimension of random fractals. New York J. Math., 7, 99–115.

Edgar, G. A. (1998) Integral, Probability and Fractal Measures. Springer-Verlag New York.

Edgar, G. A. (2008) Measures, Topology and Fractal Geometry. Undergraduate texts in Maths. Undergraduate texts in mathematics, Springer-Verlag.

Eggleston, H. G. (1949) The fractional dimension of a set defined by decimal properties. J. Math., 20, 31–36.

Eggleston, H. G. (1951) The fractional dimension of a set defined by decimal properties. J. Math., 20, 31–36.

Falconer, K. J. (1994) The multifractal spectrum of statistically self-similar measures. J. Theor. Probab., 7(3), 681–702.

Falconer, K. J. (1985) The geometry of fractal sets. Lecturer in Mathematics, University of Bristol.

Falconer, K. J. (1990) Fractal Geometry. John Wiley and Sons.

Kenneth, J. Falconer (auth.), Christoph Bandt, Siegfried Graf, Martina Zähle (1995) Fractal Geometry and Stochastics, Series: Progress in Probability, Volume 37, Birkhäuser Basel, Boston.

Fan, A. H. and Feng, D. J. (1998) Analyse multifractale de la récurence sur l'espace symbolique. CRAS. Paris, t. 327, Série, 1, 629–632.

Fan, Q., Liu, S. and Wang, K. (2019) Multiscale multifractal detrended fluctuation analysis of multivariate time series. Physica A, 532, 121864. DOI: 10.1016/j.physa.2019.121864.

Fernandez-Martinez, M., Guirao, J. L. G., Sanchez-Granero, M. A. and Segovia, J. E. T. (2019) Fractal Dimension for Fractal Structures: With Applications to Finance. SEMA SIMAI Springer Series 19, Springer International Publishing. ISBN: 978-3-030-16644-1;978-3-030-16645-8.

de Figueiredo, B. C. L., Moreira, G. R., Stosic, B. and Stosic, T. (2014) Multifractal analysis of hourly wind speed records in petrolina, northeast brazil. Rev. Bras. Biom., Sao Paulo, 32(4), 599–608.

Fillol, J. (2005) Modélisation multifractale du taux de change dollar/euro. Economie Internationale, (104), 135–150.

Flake, G. W. (1998) The Computational Beauty of Nature: Computer Explorations of Fractals, Chaos, Complex Systems, and Adaptation. MIT Press (MA).

Frostman, O. (1935) Potentiel d'équilibre et capacité des ensembles avec quelques applications à la théorie des fonctions. Thèse de Doctorat de la Faculté des Sciences de Lund.

Frame, M. and Cohen, N. (2015) Fractals and Dynamics in Mathematics, Science, and the Arts: Theory and Applications Benoit Mandelbrot: A Life in Many Dimensions, World Scientific.

Frisch, U. and Parisi, G. (1985) Fully developed turbulence and intermittency, Proc. Int. Summer school Phys., Enrico Fermi, North Holland, 84–88.

Garcia, I. and Zuberman, L. (2012) Exact packing measure of central Cantor sets in the line. Journal of Mathematical Analysis and Applications, 386(2), 801–812.

Garcia, I., Molter, U. M. and Scotto, R. (2007) Dimension functions of Cantor sets. Proc. Am. Math. Soc., 135(10), 3151–3161.

Gatzouras, D. (1999) Lacunarity of self-similar and stochastically self-similar sets. Trans of AMS, 352(5), 1953–1983.

Gervais, J.-J. (2009) Introduction aux fractals et aux systèmes dynamiques. 2e version, Août.

Guo, Q. Jiang, H. and Xi, L. (2006) Hausdorff Dimension of Generalized Sierpinski Carpet. Interna. J. Nonlinear Sci, 6, 153–158.

Gurevich, B. M. and Tempelman, A. A. (1999) Hausdorff dimension and thermodynamic formalism. Communication of the Moscow Mathematical Society.

Guzman, M. (1975) Differentiation of Integrals on $\mathbb{R}^d$. Lecture Notes in Mathematics. N 481. Springer Verlag.

Halsey, T. C., Jensen, M. H., Kadano, L. P., Procaccia, I. and Shraiman, B. I. (1986) Fractal measures and their singularities: The characterization of strange sets. Phys. Rev. A, 33, 1141–1151.

Hare, K., Mendivil, F. and Zuberman, L. (2013) The sizes of rearrangements of Cantor sets. Cana. Math. Bull., 56(2), 354–365.

Hare, K. and Zuberman, L. (2010) Classifying Cantor sets by their multifractal spectrum. Nonlinearity, 23(11), 2919–2933.

Heurteaux, Y. (1998) Estimations de la dimension inférieure et de la dimension supérieure des mesures. Ann. Ins. H. Poicaré. Probab. Statist., 34, 309–335.

Huang, X. and El-Sayed, M. A. (2010) Gold nanoparticles: Optical properties and implementations in cancer diagnosis and photothermal therapy. Journal of Advanced Research, 1(1), 13–28.

Hudson, R. L. and Mandelbrot, B. B. (2005) The Misbehavior of Markets: A Fractal View of Financial Turbulence: A Fractal View of Risk, Ruin, and Reward, Basic Books.

Hutchinson, J. (1981) Fractals and self-similarity. Indiana Univ. Math. J., 30, 713–747.

Jarvenpaa, E., Jarvenpaa, M., Kaenmaki, A., Rajala, T., Rogovin, S. and Suomala, V. (2017) Packing Dimension and Ahlfors Regularity of Porus Sets in Metric Spaces. arXiv:1701.08593v1 [math.CA] 30 Jan 2017.

Karaca, Y. and Cattani, C. (2017) Clustering multiple sclerosis subgroups with multifractal methods and self-organizing map algorithm. Fractals, 25(04), 1740001.

Karaca, Y., Cattani, C., Moonis, M. and Bayrak, S. (2018) Stroke subtype clustering by multifractal bayesian denoising with fuzzy means and-means algorithms. Complexity.

Khalili, G. A. and Cattani, C. (2019) Fractal logistic equation. Fractal and Fractional, 3(3), 41.

Kumar, S., Nisar, K. S., Kumar, R. Cattani, C. and Samet, B. (2020) A new Rabotnov fractional-exponential function-based fractional derivative for diffusion equation under external force. Math. Meth. Appl. Sci., 43(7), 1–12.

King, J. (1995). The singularity spectrum for general Sierpinski carpets. Adv. Math., 116, 1–11.

Kigami, J. (2001). Analysis on Fractals. Cambridge Tracks in Mathematics. Cambridge University Press.

Lambert, A. (2012) Théorie de la mesure et intégration, Exercices et Corrigés, Licence de Mathématiques L3, UE LM364 & LM365, Université Pierre et Marie Curie (Paris 6), Année 2012–2013.

Li, M., Zhao, W. and Cattani, C. (2013) Delay bound: fractal traffic passes through network servers. Mathematical Problems in Engineering.

Liu, K., Hu, R. J., Cattani, C., Xie, G. N., Yang, X. J. and Zhao, Y. (2014) Local Fractional-Transforms with Applications to Signals on Cantor Sets. Abstract and Applied Analysis.

Li, M., Hui, X. F., Cattani, C., Yang, X. J. and Zhao, Y. (2014) Approximate solutions for local fractional linear transport equations arising in fractal porous media. Advances in Mathematical Physics.

Liu, S., Cattani, C. and Zhang, Y. (2019) Introduction of fractal based information processing and recognition. Appl. Sci., 9(7), 1297.

Lucas, A. (2000) Mesure de Hausdorff d'ensembles liés aux oscillations de Wienner. CRAS. Paris, t. 330, Série 1, 509–512.

Makarov, N. G. (1985). On the distortion of boundary sets under conformal mappings. Proc. Lond. Math. Soc., 51, 369–384.

Mandelbrot, B. B. (1974) Intermittent turbulence in self-similar cascades divergence of high moments and dimension of the carrier. J. Fluid Mech., 62, 331–358.

Mandelbrot, B. B. (1982) The Fractal Geometry of Nature, 1st edn., W. H. Freeman and Company.

Mandelbrot, B. B. (1993). Negative dimensions and Hölders, multifractals, and the role of lateral preasymptotics in science. Fourier Analysis and Applications, Paris. Edited by J. Peyrière et J. P. Kahane meeting.

Mandelbrot, B. and Ried, R. (1995). Multifractal formalism for infinite multinomial measures. Adv. Appl. Math., 16, 132–150.

Mandelbrot, B. B. (1997) Fractals and Scaling in Finance. Springer.

Mandelbrot, B. B. (1999) Les objets fractals: forme, hasard et dimension, survol du langage fractal, 4th edn., Flammarion.

Mandelbrot, B. B. (2004) Fractals and Chaos: The Mandelbrot Set and Beyond, 1st ed., Springer-Verlag New York.

Mandelbrot, B. B. and Hudson, R. L. (2006) The Misbehavior of Markets: A Fractal View of Financial Turbulence, Basic Books.

Mandelbrot, B. B. (2012) The Fractalist, Memoir of a Scientific Maverick Pantheon.

Meneveau, C., Sreenivasan, K. R., Kailasnath, P. and Fan, M. S. (1990) Joint multifractal measures: Theory and applications to turbulence. Physical Review A, 41(2), 894–913.

Mattila, P. (1995) Geometry of Sets and Measures in Euclidean Spaces: Fractals and Rectifiability. Cambridge University Press.

Mauroy, B., Filoche, M., Weibel, E. R. and Sapoval, B. (2004) An optimal bronchial tree may be dangerous. Nature, 427, 633–636.

Mearns, B. (2021) Fractal Fern (https://www.mathworks.com/matlabcentral/fileexchange/4372-fractal-fern), MATLAB Central File Exchange. Retrieved April 12, 2021.

Menceur, M., Ben Mabrouk, A. and Betina, K. (2016) The multifractal formalism for measures, review and extension to mixed cases. Anal. Theory Appl., 32(1), 77–106.

Menceur, M. and Ben Mabrouk, A. (2019) A joint multifractal analysis of vector valued non Gibbs measures. Chaos, Solitons Fractals, 126, 203–217.

Mignot, P. (1998) Une méthode d'estimation du spectre multifractal, CRAS. Paris, 327(I), 689–692.

Ngai, S. M. (1997) A dimension result arising from the L^q-spectrum of a measure. Proc. AMS, 125(10), 2943–2951.

Nguyen, N. (2001) Iterated function systems of finite type and week separation property. Proc. AMS., 130(2), 483–487.

Novak, M. M. (2004) Thinking in Patterns: Fractals and Related Phenomena in Nature. World Scientific.

Olivier, E. (1998) Analyse multifractale de fonctions continues, CRAS. Paris, 326(I), 1171–1174.

Olsen, L. (1995) A multifractal formalism. Adv. Math., 116, 82–196.

Olsen, L. (1996) Multifractal dimensions of product measures. Math. Proc. Camb. Phil. Soc., 120, 709–34.

Olsen, L. (2000) Dimension inequalities of multifractal Hausdorff measures and multifractal packing measures. Math Scand, 86, 109–29.

Olsen, L. (2003) Mixed divergence points of self-similar measures. Indiana Univ. Math. J., 52, 1343–1372.

Olsen, L. (2004) Mixed generalized dimensions of self-similar measures. J. Math. Anal. Appl., 306, 516–539.

Olsen, L. (auth.), Christoph Bandt, Siegfried Graf, Martina Zähle (2000) Fractal Geometry and Stochastics II, Series: Progress in Probability (ed.), Vol. 46, Birkhäuser Basel, Boston.

O'Neil, T. C. (1997) The multifractal spectrum of quasi self-similar measures. J. Math. Anal. Appl., 211, 233–257.

Patzschke, N. (1997) The strong open set condition in the random case. Proc. AMS., 125(7), 2119–2125.

Pesin, Y. B. (1997) Dimension Theory in Dynamical Systems. Contemporary Views and Applications. Chicago Lectures in Mathematics.

Pesin, Y. B. and Climenhaga, V. (2009) Lectures on Fractal Geometry and Dynamical Systems. Students Mathematical Library, Vol. 52, Mathematics Advanced Study, American Mathematical Society.

Peitgen, H.-O., Saupe, D., Fisher, Y., McGuire, M., Voss, R. F., Barnsley, M. F., Devaney, R. L. and Mandelbrot, B. B. (1988) The science of fractal images, 1st edn., Springer-Verlag.

Peitgen, H.-O., Jurgens, H. and Saupe, D. (1992), Chaos and Fractals. New Frontiers of Science. Springer Verlag.

Peitgen, H.-O., Jurgens, H. and Saupe, D. (1992) Fractals for the Classroom: Part Two: Complex Systems and Mandelbrot Set, 1st edn., Springer-Verlag, New York.

Peyrière, J. (1992) Multifractal measures. Dordrecht Kluwer Academic Press, ed Proceeding of the NATO advanced study institute on probabilistic and stochastic methods in analysis with applications, Vol. 372.

Pickover, C. A. (1995) The Pattern Book: Fractals, Art, and Nature. World Scientific.

Rashid, T. (2014) Make Your Own Mandelbrot, CreateSpace Independent Publishing Platform.

Rand, D. A. (1989) The singularity spectrum $f(\alpha)$ for Cookie-Cutters, Ergod. Th. and Dynam. Sys., 9, 527–541.

Raymond, X. S. and Tricot, C. (1988) Packing regularity of sets in n-space. Math. Proc. Camb. Phil. Soc., 103, 133–145.

Redondo, J. M. (1993) Fractal models of density interfaces. IMA. Conf. Ser. 13, (ed. M. Frage, JCR, Hunt, JC Vassilicos), 353–370, Elsevier.

Riedi, R. (1995) An improved multifractal formalism and self-similar measures. J. Math. Anal. Appl., 189, 527–541.

Riedi, R. H. and Scheuring, I. (1997) Conditional and relative multifractal spectra. Fractals: An Interdisciplinary J., 5, 153–68.

Robert, L.-H. (2020) Théorie de la mesure et intégration. Université de Genève, Printemps, Section de Mathématiques.

Rogers, C. A. (1970) Hausdorff Measures, Cambridge University Press.

Sapoval, B. and Filoche, M. (2010) Pourquoi le poumon est-il aussi robuste? La Recherche, Février.

Scheuring, I. and Riedi, R. H. (1994) Application of multifractals to the analysis of vegetation pattern. Journal of Vegetation Science, 5, 489–496.

Scholz, C. H. and Mandelbrot, B. B. (1989) Fractals in Geophysics. In Pure and Applied Geophysics, 1 ed., Birkhäuser Basel.

Selezneff, A. (2011) Critères de capacité nulle. Mémoire de Maîtrise en Mathématiques, Département de Mathématiques et de Statistique, Faculté des Sciences et de Génie, Université Laval, Québec.

Sierpinski, W. (1916) Sur une courbe cantorienne qui contient une image biunivoque et continue de toute courbe donnée. C. R. Acad. Sci. Paris (in French). 162, 629–632.

Spear, D. W. (1992) Measures and self-similarity. Adv. Math., 91, 143–157.

Stauffer, D. and Stanley, E. H. (1991) From Newton to Mandelbrot: A Primer in Modern Theoretical Physics, 2nd edn., Springer-Verlag.

Taylor, S. J. (1995) The fractal analysis of Borel measures in $\mathbb{R}^d$. J. Fourier. Anal. Appl., Kahane Special Issue, 553–568.

Triebel, H. (1992) Theory of function spaces II. Birkhäuser Verlag. Basel. Boston, Berlin.

Veerman, P. (1998) Hausdorff Dimension of Boundaries of Self-Affine Tiles in $\mathbb{R}^n$. Bol. Mex. Mat, 34, 1–24.

Vehel, J. L. and Legrand, P. (2004) Signal and image processing with FRACLAB, DOI: 10.1142/9789812702746_0032, FracLab, Available http://www-rocq. inria.fr/fractales.

Vehel, J. L. and Vojak, R. (1998) Multifractal analysis of Choquet capacities. Adv. Appl. Math., 20, 1–43.

Walter, C. (2001), Les Echelles de Temps sur les Marchés Financiers. Revue de Synthèse, 4(1), 55–69.

Wang, L. F., Yang, X. J., Baleanu, D., Cattani, C. and Zhao, Y. (2014) Fractal dynamical model of vehicular traffic flow within the local fractional conservation laws. Abstract and Applied Analysis.

Wanqing, S., Chen, X., Cattani, C. and Zio, E. (2020) Multifractional Brownian Motion and Quantum-Behaved Partial Swarm Optimization for Bearing Degradation Forecasting. Complexity.

Weibel, E. R. (1963) Morphology of the Human Lung, Springer Verlag and Academic Press, Heidelberg-New York.

Wu, J. M. (1998) Hausdorff dimension on metric spaces. Proc. AMS, 126(5), 1453–1459.

Wu, M. (2005) The multifractal spectrum of some Moran measures. Science in China Series A: Mathematics Volume, 48, 1097–1112. Doi: 10.1360/ 022004-10.

Wu, M. (2005) The singularity spectrum $f(\alpha)$ of some Moran fractals. Monatsh Math, 144, 141–155.

Wu, M. and Xiao, J. (2011) The singularity spectrum of some non-regularity Moran fractals. Chaos Solitons Fractals, 44, 548–557.

Xiao, J. and Wu, M. (2008) The multifractal dimension functions of homogeneous Moran measure. Fractals, 16, 175–185.

Xiaojun, Y., Zhao, Y., Baleanu, D., Cattani, C. and Cheng, D. F. (2013) Maxwell's Equations on Cantor Sets: A Local Fractional Approach.

Xu, S., Ling, X., Cattani, C., Xie, G. N., Yang, X. J. and Zhao, Y. (2014) Local fractional Laplace variational iteration method for nonhomogeneous heat equations arising in fractal heat flow. Mathematical Problems in Engineering.

Xu, M. and Wang, S. (2011) The boundedness of bilinear singular integral operators on Sierpinski Gaskets. Anal. Theory Appl., 27, 92–100.

Xu, S. and Xu, W. (2012) Note on the Paper, An Negative Answer to a Conjecture on the Self-similar Sets Satisfying the Open Set Condition. Anal. Theory Appl., 28, 49–57.

Xu, S., Xu, W. and Zhong, D. (2012) Some new iterated function systems consisting of generalized contractive mappings. Anal. Theory Appl., 28, 269–277.

Yang, A. M., Cattani, C., Zhang, C., Xie, G. N. and Yang, X. J. (2014) Local fractional Fourier series solutions for nonhomogeneous heat equations arising in fractal heat flow with local fractional derivative. Advances in Mechanical Engineering, 6, 514639.

Yang, X. J., Cattani, C. and Xie, G. (2015) Local fractional calculus application to differential equations arising in fractal heat transfer. Fract. Dyn., 272–285.

Yang, X. J., Machado, J. A. T., Cattani, C. and Gao, F. (2017) On a fractal LC-electric circuit modeled by local fractional calculus. Commun. Nonlinear Sci. Numer. Simul., 47, 200–206.

Yang, X. J., Srivastava, H. M. and Cattani, C. (2015) Local fractional homotopy perturbation method for solving fractal partial differential equations arising in mathematical physics. Rom. Rep. Phys., 67(3), 752–761.

Yang, A. M., Zhang, Y. Z., Cattani, C., Xie, G. N., Rashidi, M. M., Zhou, Y. J. and Yang, X. J. (2014) Application of local fractional series expansion method to solve Klein-Gordon equations on Cantor sets. Abstr. Appl. Anal., 55.

Yang, A. M., Zhang, C., Jafari, H., Cattani, C. and Jiao, J. (2014) Picard successive approximation method for solving differential equations arising in fractal heat transfer with local fractional derivative. Abstr. Appl. Anal., 2014.

Yan, S. H., Chen, X. H., Xie, G. N., Cattani, C. and Yang, X. J. (2014) Solving Fokker-Planck equations on Cantor sets using local fractional decomposition method. Abstr. Appl. Anal.

Ye, Y.-L. (2007) Self-similar vector-valued measures. Adv. Appl. Math., 38, 71–96.

Yeh, J. (2014) Real Analysis: Theory of Measure and Integration. World Scientific.

Yu, Q., Dave, R. N., Zhu, C., Quevedo, J. A. and Pfeffer, R. (2005) Enhanced fluidization of nanoparticles in an oscillating magnetic field. AIChE J., 51(7), 1971–1979.

Yuan, Y. (2015) Spectral self-affine measures on the generalized three sierpinski gasket. Anal. Theory Appl., 31, 394–406.

Zhang (2018) Physical fundamentals of nanomaterials. Micro & Nano Technologies Series, Chemical Industry Press, William Andrew Applied Sciences Publisher. Elsevier.

Zhang, Y., Cattani, C. and Yang, X. J. (2015) Local fractional homotopy perturbation method for solving non-homogeneous heat conduction equations in fractal domains. Entropy, 17(10), 6753–6764.

Zeng, C., Yuan, D. and Xui, S. (2012) The hausdorff measure of sierpinski carpets basing on regular pentagon. Anal. Theory Appl., 28, 27–37.

Zhou, Z. and Feng, L. (2012) A theoretical framework for the calculation of hausdorff measure self-similar set satisfying OSC. Anal. Theory Appl., 27, 387–398.

Zhu, Z. and Zhou, Z. (2014) A local property of hausdorff centered measure of self-similar sets. Anal. Theory Appl., 30, 164–172.

Index

215

Series on Advances in Mathematics for Applied Sciences

Editorial Board

Series on Advances in Mathematics for Applied Sciences

Aims and Scope

This Series reports on new developments in mathematical research relating to methods, qualitative and numerical analysis, mathematical modeling in the applied and the technological sciences. Contributions rlated to constitutive theories, fluid dynamics, kinetic and transport theories, solid mechanics, system theory and mathematical methods for the applications are welcomed.

This Series includes books, lecture notes, proceedings, collections of research papers. Monograph collections on specialized topics of current interest are particularly encouraged. Both the proceedings and monograph collections will generally be edited by a Guest editor.

High quality, novelty of the content and potential for the applications to modern problems in applied science will be the guidelines for the selection of the content of this series.

Instructions for Authors

Submission of proposals should be addressed to the editors-in-charge or to any member of the editorial board. In the latter, the authors should also notify the proposal to one of the editors-in-charge. Acceptance of books and lecture notes will generally be based on the description of the general content and scope of the book or lecture notes as well as on sample of the parts judged to be more significantly by the authors.

Acceptance of proceedings will be based on relevance of the topics and of the lecturers contributing to the volume.

Acceptance of monograph collections will be based on relevance of the subject and of the authors contributing to the volume.

Authors are urged, in order to avoid re-typing, not to begin the final preparation of the text until they received the publisher's guidelines. They will receive from World Scientific the instructions for preparing camera-ready manuscript.

Series on Advances in Mathematics for Applied Sciences

www.ingramcontent.com/pod-product-compliance
Lightning Source LLC
Chambersburg PA
CBHW050556190326
41458CB00007B/2068